SpringerBriefs in Molecular Science

Chemistry of Foods

Series Editor

Salvatore Parisi, Al-Balqa Applied University, Al-Salt, Jordan

The series Springer Briefs in Molecular Science: Chemistry of Foods presents compact topical volumes in the area of food chemistry. The series has a clear focus on the chemistry and chemical aspects of foods, topics such as the physics or biology of foods are not part of its scope. The Briefs volumes in the series aim at presenting chemical background information or an introduction and clear-cut overview on the chemistry related to specific topics in this area. Typical topics thus include:

– Compound classes in foods—their chemistry and properties with respect to the foods (e.g. sugars, proteins, fats, minerals, …)
– Contaminants and additives in foods—their chemistry and chemical transformations
– Chemical analysis and monitoring of foods
– Chemical transformations in foods, evolution and alterations of chemicals in foods, interactions between food and its packaging materials, chemical aspects of the food production processes
– Chemistry and the food industry—from safety protocols to modern food production

The treated subjects will particularly appeal to professionals and researchers concerned with food chemistry. Many volume topics address professionals and current problems in the food industry, but will also be interesting for readers generally concerned with the chemistry of foods. With the unique format and character of SpringerBriefs (50 to 125 pages), the volumes are compact and easily digestible. Briefs allow authors to present their ideas and readers to absorb them with minimal time investment. Briefs will be published as part of Springer's eBook collection, with millions of users worldwide. In addition, Briefs will be available for individual print and electronic purchase. Briefs are characterized by fast, global electronic dissemination, standard publishing contracts, easy-to-use manuscript preparation and formatting guidelines, and expedited production schedules.

Both solicited and unsolicited manuscripts focusing on food chemistry are considered for publication in this series. Submitted manuscripts will be reviewed and decided by the series editor, Prof. Dr. Salvatore Parisi.

To submit a proposal or request further information, please contact Dr. Sofia Costa, Publishing Editor, via sofia.costa@springer.com or Prof. Dr. Salvatore Parisi, Book Series Editor, via drparisi@inwind.it or drsalparisi5@gmail.com

More information about this subseries at http://www.springer.com/series/11853

Candela Iommi

Chemistry and Safety of South American *Yerba Mate* Teas

Springer

Candela Iommi
Food Safety and Public Health Consultant
Milan, Italy

ISSN 2191-5407   ISSN 2191-5415 (electronic)
SpringerBriefs in Molecular Science
ISSN 2199-689X   ISSN 2199-7209 (electronic)
Chemistry of Foods
ISBN 978-3-030-69613-9   ISBN 978-3-030-69614-6 (eBook)
https://doi.org/10.1007/978-3-030-69614-6

This Springer imprint is published by the registered company Springer Nature Switzerland AG
The registered company address is: Gewerbestrasse 11, 6330 Cham, Switzerland

Contents

Chapter 1
Yerba Mate Tea, a Traditional South American Beverage. An Introduction

Abstract This chapter concerns the traditional use of a peculiar plant—*Ilex paraguariensis*—diffused in South America, and widely present in four countries of this area: Argentina, Uruguay, Paraguay, and Brazil. This plant is used for centuries because of different claimed properties, in form of beverage (infusion), under the following names: *mate, chimarrão*, and *tererê*. The interest in these infusions depends on the global diffusion of these products while the origin of *I. paraguariensis* is in the South American area. On the one side, this plant and his infusions are known because of different pharmacological effects against common illnesses. On the other side, different processing methods mainly based on drying and grinding steps are able to modify notably sensorial and chemical properties of sold products. The analysis of historical traditions linked with *yerba mate* can be useful when speaking of different topics including regional or national differences, authenticity and traceability issues on a global scale. Consequently, a deep analysis of the current market and new marketing perspectives should be recommended.

Keywords Authenticity · *Ilex paraguariensis* · Cultural heritage · *Yerba mate* · *Chimarrão* · South America · Polyphenols · *Tererê*

Abbreviations

EU	European Union
FNCF	Federazione Nazionale dei Chimici e dei Fisici
FBO	Food Business Operator
GI	Geographical Origin
INYM	*Instituto Nacional de la Yerba mate*
MERCOSUR	*Mercado Común del Sur*
MSME	Ministry of Micro, Small & Medium Enterprises
RASFF	Rapid Alert System for Food and Feed
EPA	United States Environmental Protection Agency

© The Author(s), under exclusive license to Springer Nature Switzerland AG 2021
C. Iommi, *Chemistry and Safety of South American Yerba Mate Teas*,
Chemistry of Foods, https://doi.org/10.1007/978-3-030-69614-6_1

1.1   Vegetable Phenolics for Food Purposes. A General Introduction

The matter of phenolic compounds in the food and beverage industry is extremely interesting, and the related perception has constantly increased in recent years. The research of terms such as 'phenolic' associated with 'plant' and 'food' on two online scholar platforms (https://www.crossref.org and https://scholar.google.com) can give remarkable results. In particular, Figs. 1.1 and 1.2 show the amount and the increasing trend of journal papers in Crossref.org and Scholar Google, respectively, with the above-mentioned words, between 2011 and 2020. Figure 1.1 shows that the increasing number of papers in this ambit is constant (only the 2020-number is slightly lower than in 2011), ranging from 43,534 papers in 2011 to 76,794 items in 2020. Figure 1.2 shows a similar trend, even if the increase of related journal papers is less pronounced (Table 1.1).

The interest in polyphenols is essentially correlated with safety and hygiene features of many vegetable and fruit-based foods and beverages because of their chemical properties, in particular antioxidant features (Barbera 2020; Barbieri et al. 2019; Bhagat et al. 2019; Delgado et al. 2016, 2017, 2019; Parisi 2019, 2020; Singla et al. 2019). However, the list of topics of interest is really long, when speaking of active principles with hygiene and health features of vegetable origin (Chammem et al. 2018; Camargo and Schwember 2019; Toffoli et al. 2019; Dibanda et al. 2020; Efing et al. 2009; Fiorino et al. 2019; Flamminii et al. 2019; Folch 2010; Garcia-Lazaro et al. 2020; Goeddel 2015; Goldenberg et al. 2003; Goyke 2017; Greizerstein et al. 2004; Habtemariam 2019; Haddad et al. 2020a, b; Issaoui et al. 2020; Panzella et al. 2020; Parisi 2012, 2013; Parisi and Dongo 2020; Parisi et al. 2020; Shahidi et al. 2019; Srivastava 2019):

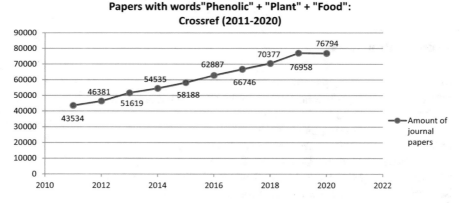

Fig. 1.1 This graph shows the increasing trend of the amount of scientific papers containing the following words: 'phenolic', 'plant', and 'food' on the online scholar platform https://www.crossref.org (period: 2011 to 2020). Only the 2020-number is slightly lower than in 2011, ranging from 43,534 papers in 2011 to 76,794 items in 2020

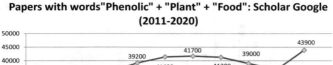

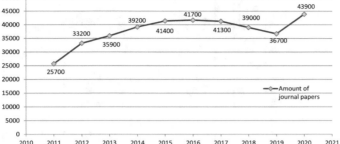

Fig. 1.2 This graph shows the increasing trend of the amount of scientific papers containing the following words: 'phenolic', 'plant', and 'food' on the online scholar platform https://scholar.goo gle.com. (period: 2011 to 2020). In general, the increase of related journal papers is less pronounced if compared with https://www.crossref.org (Fig. 1.1). The 2020-result is significantly increased if compared with the previous years, in particular with respect to 2019

(1)   Sustainable agricultural practices
(2)   Remediation procedures and sensorial profiles of wastewaters from food industries containing polyphenols
(3)   Technological strategies concerning the design of new 'functional foods', also including the possible use of polyphenols as food additives (with natural claims)
(4)   Analytical chemistry, with peculiar relation to laboratory procedures
(5)   Effect of food processing on antioxidant properties, with some consideration concerning the use of food packaging solutions and the development of active and intelligent packages
(6)   Historical variations of vegetable and fruit characters along the human history, with considerations concerning the cultural heritage of certain populations in the Mediterranean Area, the Middle East, the South American Countries, etc.
(7)   Shelf-life issues depending on the availability of polyphenols and other antioxidant principles in certain foods and beverages, taking into account the Parisi's first Law of Food Degradation.

A remarkable part of the scientific literature concerning polyphenols and other active principles of safety and health interest concerns the abundance of these compounds in several well-known vegetable species commonly found in Europe, the Middle East, or the Far East. However, a relevant amount of information remains to be discovered and researched when speaking of other world areas with a notable heritage. In particular, the aim of this chapter is to introduce the reader in the ambit of vegetable infusions commonly found and consumer in the Latin America areas (Janda et al. 2020; Magri et al. 2020; Matta 2019; Olivari et al. 2020).

Table 1.1 A matrix of correlation between basic *yerba mate* typologies, from the commercial viewpoint, and seven technological parameters

Quality	Selection of leaves?	Exclusively from natural plantations?	Concomitant *barbaqua/secado* process ?	Particle sizes	Amount of twigs	Roasting process	Genetic variations
Premium	Yes	Yes	No	–	–	No	No
Nativa	No	Yes	No	–	–	No	No
Traditional	No	No	Yes	–	–	No	No
Moida grossa	No	No	Yes	>*Moida grossa*	–	No	No
Exportação	No	No	Yes	–	Reduced	Yes	No
Tererê	No	No	Yes	Approx. 500 μm	–	No	No
Tostada	No	No	Yes	Notable	–	Yes	No
Cambona	–	–	–	–	–	–	Yes

1.2 Tea Infusions and Regional Products. A South American Perspective

In general, tea infusions are a successful product worldwide, both for historical reasons and for safety-related claims, when speaking of human health. The tradition of tea infusion originated in the Far East (China) at least 2000 years ago as a popular non-alcoholic beverage, and its presence is widespread on condition that the main raw material—tea plant leaves—is present. Actually, tea plants (*Camellia assamica, cambodiensis*, and *sinensis*) are cultivated in tropical and sub-tropical regions worldwide. Consequently, because of the need of adequate rainfall, soil drainage, and acidity values, the four tea species knows so far—black, *oolong*, green, and brick teas—are available largely within specified geographical boundaries (Karak and Bhagat 2010; Wight 1962).

Actually, tea infusions are not the only possible non-alcoholic beverage. However, these products are one of the first beverage categories worldwide with relation to new launches, excluding wines, spirits, and other alcohol-based items (Sloan and Adam Hutt 2012). Another obvious choice is represented by coffee products, at present, while different infusions such as *kawa* products remain in the so-called area of niche foods. The predominant position of coffee and tea products appears solid and should remain unchanged in the next future.

Things may change when speaking of health and safety: while coffee derivatives are not generally associated with safety advantages for the common consumer, tea products are associated with health because of their interesting active principles. Safety and health advantages are often claimed against cardiovascular illnesses, neurodegenerative disorders, cancer, obesity, lung problems, type-2 diabetes, pancreatitis, etc. In general, all these effects (including positive metabolism and weight regulation in human beings) are ascribed to polyphenols contained in vegetable infusions. Consequently, the research has been oriented to the study of known tea varieties, and also other products which might give similar results.

In a regional perspective, the Far East is generally linked to tea products, and the same thing may be affirmed when speaking of India (the second world producer after China). Europe and other continents are importing agents when speaking of tea products, in general. The South American continent is not an exception, in this ambit. However, biodiversity features in this continental area and the cultural heritage of ancient human tribes are responsible for the cultivation and the use of peculiar vegetable products which can be exported in other continents: *yerba mate*, coffee and *açaí* berries (the last fruit is not consumed as infusion). These vegetable products are very common in Brazil and in other South American Countries. In detail:

(a) *Yerba mate*, scientific name: *Ilex paraguariensis, Aquifoliaceae* family, is found naturally, in the following South American Countries: Brazil, Argentina, Paraguay, and Uruguay. It is a sub-tropical dioecious evergreen tree. For this reason, it is naturally in the eastern centre area of the South American continent. Consequently, it is not reported in the northern or southern areas of the same continent (Grigioni et al. 2004; Heck and Mejia 2007; Small and Catling

2001). In addition, the cultivation of *yerba mate* is a business, but specific and sometimes non-uniform procedures are carried out in different Countries (Giberti 1994; Heck and De Mejia2007). Historically, *yerba mate* leaves are known and appreciated as infusion since pre-Columbian era (Bracesco et al. 2011)

(b)   Coffee is broadly consumed worldwide in various forms (Matta 2019). Two main types are known and used: *Coffea arabica* and *C. canephora* (also named *C. robusta*). Actually, coffee is not a specialised tradition of South American Countries. On the contrary, coffee traditions spread historically from the ancient Ottoman Empire, specifically the Yemen region. Subsequently, coffee has been one of the distinctive symbols of colonialism: the French, Spanish, Dutch, British, Portuguese, German, Italian, and North American possessions worldwide have known coffee during the last centuries. However, coffee has specifically been cultivated in the Latin American countries: the amount of South American coffee was more than 2:1 if compared with the resting world production (based generally in Africa and Asia) in 1980. Subsequently, the non-South American production, especially in Africa, has obtained notable results: however, a typical separation between *robusta* type (grown specifically in African and some Asian Countries) and the *arabica* product (a typical South American food) persists at present (Clarence-Smith and Topik 2003)

(c)   Finally, the *açaí* berry (*Euterpe oleracea*) is an interesting case. Basically, it is a native fruit of the Amazonian area (Brazil only). It is consumed essentially as it is (fruit pulp), although different derived products can be found on the market: juices, tablets, etc. (Matta 2019). Recently, it has been reported as a product containing notable amounts of antioxidants, with consequent antioxidant activity (da Silveira et al. 2017; Matta 2019; Yamaguchi et al. 2015). This example demonstrates the economic importance of a food product which is found only in one region (the Amazonian area) and one Country (Brazil), at present.

On these bases, it can be understood that the research for new healthy foods with some regional feature and possibly a well-defined 'localisation' of the product is one of the new routes currently followed by the food and beverage industry. This situation is also observed clearly when speaking of non-tea vegetable infusions such as *yerba mate,* a peculiar product which many aspects should be discussed for.

1.3   *Ilex Paraguariensis*, a Peculiar Plant, and *Yerba Mate*

1.3.1   *Biodiversity in South America. I. Paraguariensis and Localisations*

As above mentioned, *I. paraguariensis* is diffused in South America, and widely present in four Countries of this area: Argentina (*Norte* region), Uruguay, Paraguay, and Brazil (Southern region). This plant has been and is still used for centuries because of different claimed properties, in form of beverage (infusion of *yerba mate* leaves), under the following names: *mate, chimarrão,* and *tererê*.

The commercial and scientific interest in these plants and related infusions depend on their global diffusion, while the origin of *I. paraguariensis* is in the South American area only. Differently from coffee, and similarly to *açaí* berries, it can be assumed that *I. paraguariensis* is really a specific plant of the South American continent, explaining how certain products are the most concrete expression of biodiversity.

This plant needs subtropical environments (average temperatures: 21–22 °C, with tolerated minimum temperature of 6 °C), and 1,200 mm of yearly (and possibly constant) rainfall: it flourishes from October to November, and related fruits (dark-to-red drupes, estimated length: 4–6 mm) are produced in the March–June period (Matta 2019). Consequently, snowfalls may be tolerated enough, and this aspect is important because cultivated and native regions include mountainous also (Heck and Mejia 2007). For these reasons, the most ideal habitat for *yerba mate* is concentrated in the eastern centre portion of the South American continent, as shown in Fig. 1.3 (Maccari Junior 2005). It has to be considered that Bolivia is the central country in Fig. 1.3, but *yerba mate* is not really considered in this geographical ambit. Actually, the similarity between *I. paraguariensis* and *I. argentina* (found in the area approximately located between Santa Cruz de la Sierra, Bolivia, and Andagalá, Argentina) has caused the confusion between the two plants, with the consequent attribution of *yerba mate* as a native Bolivian plant, until recent times (Giberti 1995). Interestingly, Bolivia, Peru,

Fig. 1.3 *I. paraguariensis* needs subtropical environments (average temperatures: 21–22 °C, tolerated minimum temperature: 6 °C), and 1,200 mm of yearly (and possibly constant) rainfall. The most ideal habitat for *yerba mate* is concentrated in the centre portion of the South American continent (Heck and Mejia 2007; Maccari Junior 2005)

and Chile consumers are extremely interested in *yerba mate* (Silveira et al. 2017; López 1974).

1.3.2   Native Versus Cultivated Yerba Mate. An Introduction to National Differences

Above-discussed features have also an important consequence when speaking of cultivation and harvesting techniques. In general, three methods are reported: harvesting in natural forests (on-site harvesting); cultivation and mechanical harvesting (*mate* plantations); and finally, the mixed method (harvesting and cultivation in natural forests, including re-plantation). The first system (carried out in natural forests such as Araucaria Forest or Mixed Ombrophilous Forest) has not shown good results, from the quantitative and qualitative viewpoints (Croge et al. 2020; Marques et al. 2012; Vogt et al. 2016). For this reason, the mixed system (Portuguese: *caívas*) has been progressively chosen in the Amazonian area (Brazil), with improved yields and important biopreservation results (Chaimsohn et al. 2014). The mixed systems favour the birth of *yerba mate* cultivated areas interposed within the broader Amazonian forests (Hanisch et al. 2010). With exclusive relation to Brazil, the *Rio Grande do Sul* state hosts the main part of plantation areas at present, while non-cultivated mate is located in the *Mato Grosso do Sul, Paraná*, and *Santa Catarina* States (Matta 2019). On the other hand, *mate* plantations have been created in the *Norte* region of Argentina since 1915, with extremely interesting results from the quantitative and qualitative viewpoints. Consequently, Argentina is now the first mate producer in South America with 280,000 metric tons per year in 2010, while Brazil and Paraguay are second and third in this special ranking with approximately 51 and 13.3% of the Argentinean amount, as reported in 2019 (Heck and Mejia 2007; Matta 2019). It may be also noted that the cultivation of *yerba mate* has been attempted also in other world areas, without important successes (Ilany et al. 2010).

Before discussing the real *yerba mate* nature, related infusions, production and preparation processes, their chemical properties, and other regulatory/adulteration matters, it should be noted that the particular method of production may influence the nature of obtained *yerba mate*. In fact, it has been reported that cultivated *mate* herbs are grown under the sunlight, while native (harvested only) mate is generally obtained in native forests, and consequently shaded (not in full sunlight). Consequently, chemical profiles of these two *yerba mate* types can be really different: because of the localisation of *mate* plantations in Argentina ('sunlight' *mate*) as opposed to Brazilian and non-cultivated *mate*, the comparison could give some surprise, especially when speaking of antioxidant active principles, polyphenols. These differences are substantially well observed when speaking of sensorial attributions in terms of flavour and reduced bitterness (Matta 2019).

1.3.3 Native Versus Cultivated Yerba Mate. Harvesting and Cultivation Procedures, and National Differences

The basic differences between Brazilian and Argentinean *yerba mate* strongly depend on their production methods. Actually, the most correct definition should be 'cultivation & harvesting'; on the other hand, the simple harvesting in natural forests is still carried out, with the obvious elimination of plantation procedures.

1.3.3.1 Native *Yerba Mate*. Production Steps

With exclusive reference to 'native' *yerba mate*, the following production steps have to be considered (Fig. 1.4) (Croge et al. 2020; Maccari Junior 2005; Matta 2019):

(a) On-the-ground leaf harvesting (into natural forests, no need of plantation or re-plantation)
(b) Transport of leaves (by hand or mechanical methods) to the processing facility
(c) General classification of leaves
(d) Storage of leaves
(e) *Sapeco* step. The *sapeco* word implies the rapid exposure of stored leaves to a flame (flash heating, temperature: 400–500 °C, time: 10 s-3 min) within 24 h from harvesting procedures, with the aim of breaking inner membranes and the consequent denaturation of inner enzymes such as polyphenol oxidase.

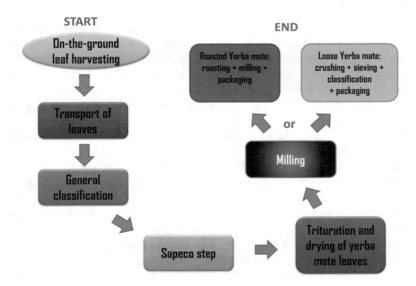

Fig. 1.4 Native *yerba mate*. The production flow chart starts from original harvesting to final roasted or loose *yerba mate* products

The treatment is critical enough because the enzymatic denaturation can avoid oxidation: in addition, moisture levels are lowered

(f)   Trituration and drying of *yerba mate* leaves. Actually, these steps can be carried out at the same time into continuous metallic cylinders, although a superior quality may be achieved if leavers are initially dried with hot air (and low thermal values are required) and subsequently triturated. The drying or *barbaqua* step (100 °C, eight to 24 h) is also named *secado*, reducing moisture to 3–6%. Other alternative versions of the same process consider 30–180 min

(g)   Milling step. It has to be noted that the raw material for this intermediate step, triturated and dried leaves, is now defined *yerba mate cancheada* (in Brazil) or *canchada* (in Argentina). The Portuguese and Spanish words *cancheada* and *canchada* mean 'thickly ground'. After this step, *yerba mate* can be obtained in two different versions (steps h or i). It has to be noted that dried leaves may be aged (9–12 months) in Argentina

(h)   Production of roasted *yerba mate*. Milled leaves are roasted (120 °C, 15 min), subjected to another milling step, and finally packed into teabag packages. This product is used for *maté cocido*: a green *yerba mate* used for producing herbal tea

(i)   Production of loose *yerba mate*. Milled leaves are crushed, sieved, classified according to grinding (coarse, medium, and fine degree), and finally: (1) packed into loose packages (commercial names: *mate, chimarrão*, and *tererê*), or (2) further aged (6–24 months) before packaging (as loose products).

Some considerations should be done here with relation to above-discussed steps (Matta 2019; Prestes et al. 2014):

(1)   The preferred harvesting period can vary. In general, Brazilian harvesters prefer to carry out the first step between May and September, with the aim of reducing plant damages. On the other side (Argentina), the preferred harvesting period should be between April and September (with one additional month if compared with Brazil). In this period, the quantity of harvested *yerba mate* leaves should approximately surpass half of the total available *I. paraguariensis* in both countries of 15%

(2)   The storage step should be considered carefully because colorimetric modifications depend not only on storage conditions but also on the time: anyway, they are irreversible and unavoidable, according to the Parisi's First Law of Food Degradation (Srivastava 2019). For this reason, Brazilian operators prefer to receive intermediate *yerba mate* leaves constantly throughout the year, with the objective of minimising colorimetric damages. In general, the most genuine *yerba mate* would be intensely green, and excessive storage times can surely modify this tint. On the contrary, Argentinean operators generally prefer an additional 9–12 month storage (aging) before milling (*yerba mate canchada*) with the aim of modifying the colorimetric appearance of green leaves and improving the typical flavour. The same operation can be carried out before packaging loose *mate* (in the final step)

(3)   With relation to the trituration and drying step (unless the drying is carried out firstly), the basic objective is to obtain a final moisture for leaves up to 3–6%. The *cancheada* leaves have to contain a reduced water amount; otherwise, non-constant moisture levels and other possible differences related to times and thermal values could affect negatively and in a cumulative way many chemical profiles of the final *yerba mate*
(4)   The roasted *yerba mate* is especially preferred in the Brazilian market
(5)   The milling step and subsequent size-reducing procedures are important enough. In Brazil, the desired particle size should not exceed 300 μm, while Argentinean operators desire higher values (5 mm).

As a consequence, there are more than one single *yerba mate* on the South American markets and worldwide. In addition, sensorial factors able to describe *yerba mate* may notably vary depending on processing steps. In general, the following parameters are the basis for sensorial description when speaking of commercial *yerba mate* (Heck and Mejia 2007):

(1)   General appearance: in terms of stick and leaf dimensions, uniformity, amount of sticks, and measurable amount of *mate* dust
(2)   General features of *mate* infusions: in terms or visible sediments, turbidimetric values, and colorimetric tint (it should be normally brownish)
(3)   Aroma and flavour features: acid taste, humid appearance, smoke aroma, green appearance, bitter taste, toasted flavour, perceivable presence of residuals.

1.3.3.2   Cultivated *Yerba Mate* (Including also Mixed Systems) Production Steps

With reference to non-native' *yerba mate*, the basic difference concerns the possibility of cultivating and re-planting existing mate plants. In the first situation, new plantation areas are needed, while the mixed system requires only that mate plants are continually replaced after the existing plants have been damaged excessively (or after death). In the ambit of cultivated and mixed productions, high yields have and are still notable if compared to 'native' (harvesting only) systems. However, the mechanisation of harvesting processes and the circumstance that monocultures are exposed to full sun may determine some metabolic alteration with visible and sensorially perceivable damages to leaves. Substantially, cultivated *yerba mate* has progressively turned into Ombrophilous species (such as natural plants living in dense forests, without full sunlight). Under the action of full sunlight, some modifications have to be considered, including also high sensibility to insects and illnesses (Marques et al. 2012). In this situation, the mixed system may favour the resistance of *I. paraguariensis*, because integrated cultivations inside forests should suffer light effects when speaking of diseases, lack of fertilisers, and plant shading/defence from full sunlight. The qualitative result is reported to be excellent, at least when speaking of flavour features. On the other hand, some important modifications of cleared soils (initially covered by forests) with *yerba mate* plantations (no mixed systems) have been reported when speaking of organic and mineral chemical profiles in the same soils (Piccolo et al. 2004).

With the exception of cultivation and/or re-plantation steps, the flow chart for non-native *yerba mate* is similar to the process for native leaves (Fig. 1.4) (Maccari Junior 2005; Matta 2019):

(a) On-the-ground leaf harvesting
(b) Transport of leaves to the processing facility
(c) General classification of leaves
(d) Storage of leaves
(e) *Sapeco* step (flash heating, temperature: 400–500 °C, time: 10 s-3 min) within 24 h from harvesting procedures
(f) Trituration and drying of *yerba mate* leaves, or initial *barbaqua* drying step (100 °C, eight to 24 h—other alternative versions of the same process consider 30–180 min) or *secado*, reducing moisture to 3–6%, followed by trituration
(g) Milling step. After this step, *yerba mate* can be obtained in two different versions (steps h or i)
(h) Production of roasted *yerba mate:* roasting (120 °C, 15 min), second milling, and finally packaging into teabag packages. This product is used for *maté cocido*: a green *yerba mate* used for producing herbal tea
(i) Production of loose *yerba mate*. Crushing, sieving, classified according to grinding (coarse, medium, and fine degree), and finally: (1) packaging into loose packages (commercial names: *mate, chimarrão*, and *tererê*), or (2) further aged (6–24 months) before packaging (as loose products).

1.4 Consumption Behaviours and *Yerba Mate* Products. One or More *Mate* Infusions? Differences Between Countries in Latin America

1.4.1 Commercial Classification, Types, and Reasons for National Diversification

As above mentioned, there is not only one *yerba mate*, but different *yerba mate* products, depending on processes, packaging choices, national or regional habits, and also the mode of consumption. Consequently, it may be difficult to define reliably safety and health effects on human beings if correlated with the consumption of *yerba mate* because of different uses. The analysis of historical traditions linked with *yerba mate* (including different names for the same product, depending on the nationality!) can be useful, because of the diffusion of the same plant in four Latin American countries (Chap. 2–6). In addition, the commerce of *mate* preparations in South America should be discussed from the regulatory viewpoint (Chap. 7). At the same time, the diffusion of *mate* teas outside of the South American area imposes some reflection when speaking of authenticity and traceability issues: the global commerce can be an advantage, but adulteration phenomena may occur at the same time (Chap. 8). Consequently, a deep analysis of the current market and new marketing perspectives should be recommended. Moreover, the cultural heritage of *yerba mate* is a notable key factor explaining partially the success of this plant worldwide.

The extreme variability of *yerba mate*-related products may be challenging enough. A first commercial classification (Fig. 1.5) can be based on the harvesting and manipulation processes (Matta 2019):

(a) *Premium*
(b) *Nativa*
(c) *Traditional,*
(d) *Moida grossa*
(e) *Exportação*
(f) *Tererê*
(g) *Tostada*
(h) *Cambona.*

These eight *yerba mate* products are differentiated on the basis of production processes and consequent commercial attributions (flavour, aroma, particle size, etc.) (Matta 2019):

(1) The *Premium* quality is obtained from first-choice selected leaves. These leaves are from natural plantations. The *secado* process is carried out reducing

Yerba mate: commercial classification

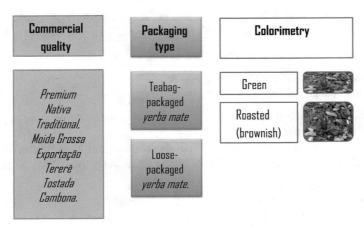

Fig. 1.5 Bases for the commercial classification of *yerba mate* types

moisture to 3–6%, followed by trituration (Sect. 1.3). The *secado* process is conducted also with the aim of avoiding excessive smoke exposure

(2) The *Nativa* type is similar to Premium quality. However, the initial first-choice selection of leaves is not carried out. As for *Premium* quality, these leaves are from natural plantations. The *secado* process is carried out reducing moisture to 3–6%, followed by trituration (Sect. 1.3). The *secado* process is conducted also with the aim of avoiding excessive smoke exposure

(3) The *Traditional* type is obtained from traditional and naturally planted leaves. Trituration and drying of *yerba mate* leaves are carried at the same time: intermediate leaves are in direct contact with produced smoke, differently from *Premium* and *Nativa* types

(4) The *Moida grossa* quality is obtained with the same *traditional* process; however, particle sizes of the final products are notable if compared with *traditional* type

(5) The *Exportação* quality is tacitly intended for export purposes. It is obtained from traditional and naturally planted leaves. Trituration and drying of *yerba mate* leaves are carried at the same time: intermediate leaves are in direct contact with produced smoke, differently from *Premium* and *Nativa* types. Subsequently, an additional eight-month storage (aging) is required at least before milling (*yerba mate canchada*) with the aim of modifying the colorimetric appearance of green leaves and improving the typical flavour. In addition, twigs are reduced as number

(6) The *Tererê* quality is obtained such as the traditional type. However, because of the destination to cold infusions, a specific particle size range is needed (approximately 500 μm because teas have a similar granulometry)

(7) The *Tostada* type is similar to *Moida grossa*. However, because of the final destinations (normal infusions, Brazilian iced teas), the product needs a roasting process

(8) Finally, the *Cambona*, variety is a genetic *I. paraguariensis* variation. The traditional process method (similarly to the *traditional yerba mate*) is required.

Another classification concerns the packaging method:

(a) Teabag-packaged *yerba mate*
(b) Loose-packaged *yerba mate.*

A third classification concerns the colorimetric appearance of leaves (green or roasted/brownish types).

On these bases, it may be inferred that the commercial classification of *yerba mate* products (Fig. 1.5) relies on the following seven variables (Matta 2019):

(1) The selection of leaves
(2) The choice between traditional or plantation method
(3) The choice of trituration and drying process, or *barbaqua/secado* procedure
(4) The selection of qualities according to particle sizes. In detail (Sect. 1.3), the grinding procedure can give coarse, medium, and fine degree qualities. A loose packaging step is required at last (commercial names: *mate, chimarrão,* and

tererê). An additional aging period (6–24 months, and also 8–12 months are reported) may be required before packaging. Consequently, green colour turns into brownish tints; flavour is enhanced)

(5)   The possible reduction of twigs (30% according to the Brazilian legislation in loose products)

(6)   The presence or absence of the roasting process (in this situation, a teabag packaging process is required at last)

(7)   The use of genetic variations of *I. paraguariensis*.

All the above-discussed variables and possible quality types are correlated together, taking into account that certain qualities may have different features. This situation may be complicated enough. However, it has to be noted that the destination of *yerba mate* products is the key for the final comprehension (Fig. 1.6). In fact, *I. paraguariensis* is mainly used (for food applications only) in the following ways (Heck and Mejia 2007):

(a)   Hot or regular infusions in water. The correspondent names are *chimarrão* or *mate*. It has to be noted that roasted *yerba mate* packaged in teabags can be used for this aim

(b)   Cold infusions in water. The correspondent name is generally *tererê*. It has to be noted that roasted *yerba mate* packaged in teabags can be used for this aim.

In addition (Fig. 1.6):

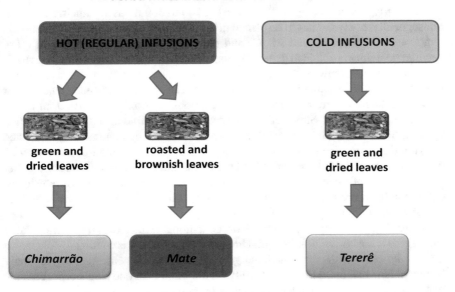

Fig. 1.6 *Yerba mate* is mainly used for hot or cold infusions in water. The correspondent names are *chimarrão* or *mate* (hot version) and *tererê* (cold infusion). *Chimarrão* and *tererê* are obtained from green and dried leaves, while *mate* derives exclusively from roasted and brownish leaves

(1)   *Chimarrão* and *tererê* are obtained from green and dried leaves
(2)   *Mate* (hot infusion) derives exclusively from roasted and brownish leaves. On these bases, a certain distinction between different *yerba mate* types can be carried out, considering also that national preferences have some importance (in general, the *mate* infusion seems to be preferred in Argentina).

A final consideration concerns the preparation method of *yerba mate* infusions because tradition and history have a notable importance, probably linked to the Precolumbian Age. After all, *yerba mate* was traditionally consumed by *Guaraní* populations, subsequently discovered and used by the Jesuits, and extremely popular after the expulsion of the Jesuits themselves (Delacassa and Bandoni 2001). The peculiar container and straw used for consumption are an artistic expression of the ancient South American history and folklore (Oberti 1960).

1.4.2   Hot Yerba Mate Infusions. Preparation Procedures (Brazil, Argentina)

The extraction of *yerba mate* substances from leaves is carried out in hot water by means of a peculiar container. In detail, and considering possible variations (Cansiani et al. 2008; Heck and Mejia 2007; Lovera et al. 2019; Pagliosa et al. 2009; Santa Cruz et al. 2002), 20–100 g of green or brownish *yerba mate* leaves (*chimarrão* or *mate*, Sect. 1.4.1) are placed into a very traditional gourd-shaped container, also named *porongo matero* (Fig. 1.7), measuring 100 to 300 ml as volumetric capacity. Subsequently, hot water has to be placed in the container (approximately 30–200 ml, temperature between 65 and 95 °C) and then sipped directly in the 'gourd'. The peculiar container is obtained generally from the calabash tree (*Lagenaria siceraria*). The visible metal straw or tube into the container, also connected to a filter on the inner bottom side, is named *bomba* or *bombilla*. The consumption ends when *yerba mate* is 'exhausted', after repeated water additions: approximately 1000 ml of hot water should be needed. Substantially, the process is a continuous infusion/extraction cycle in hot water, and the remaining *yerba mate* has to be continuously subjected to extraction during consumption 5–10 times. It should be noted that *mate* may be consumed hot or cold; however, it is obvious that hot extraction allows for a better result in terms of assumption of *I. paraguariensis* active principles. Moreover, the extraction from normal teabags containing only three *yerba mate* grams in 200 ml of hot water could not give the same results.

Interestingly, the aqueous suspension containing extracted *yerba mate* soluble substances is not hot such as the initial hot water because heat is continually dispersed through *bombilla* walls from the inner gourd to the environment. Actually, it should be recognised that *bombilla* and gourd types are more than one single type: ceramic materials, glass, and also calabash, metals or wood could be used in this ambit. In addition, used gourds may have different shapes and volumetric capacities (Lovera et al. 2019). Consequently, some variability has to be expected when speaking of hot

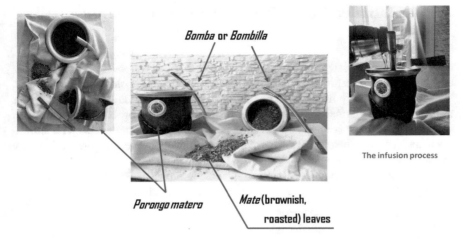

Fig. 1.7 Infusion of *yerba mate* leaves is carried out by means of a traditional gourd-shaped container, also named *porongo matero*, measuring 100 to 300 ml as volumetric capacity. Leaves are placed in the *matero*; subsequently, hot water has to be placed in the container (approximately 30–200 ml, temperature between 65 and 95 °C) and then sipped directly in the 'gourd'. The peculiar container is obtained generally from the calabash tree (*Lagenaria siceraria*). The visible metal straw or tube into the container, also connected to a filter on the inner bottom side, is named *bomba* or *bombilla*. The process is a continuous infusion/extraction cycle

yerba mate infusions because of the extreme variability of used 'gourd and *bombilla*' (or '*porongo matero y bombilla*') kits.

1.4.3   Cold Yerba Mate Infusions (Tererê). Preparation Procedures (Paraguay)

Differently from hot infusions, the yield of cold infusions (maceration in cold water) is low enough, and this conclusion seems obvious (Cansiani et al. 2008; Kujawska 2018; Kujawska and Hilgert 2014). As for *mate* and *chimarrão*, the cold infusion is consumed by means of the metallic *bombilla* or *bomba*. Consequently, there are not differences between this system, extremely popular in Paraguay and in hot days, and hot infusions, except for the temperature of extraction. Naturally, sensorial results are completely different (*tererê* could be considered as a 'fresh' beverage in hot days…).

1.5 An Introduction to Chemical Profiles of *Yerba Mate* Products

From the chemical viewpoint, the compositional profile of different *yerba mate* products influences the resulting infusion, both in the hot and in the cold versions. Anyway, because of the safety importance of these preparations as a daily consumption for million South American consumers (similarly to the growing coffee consumption worldwide), the best strategy is surely the examination of biomolecules with antioxidant and positive heath effects, provided that these compounds are sufficiently extracted in water. A complete discussion is found in Chap. 4.

By a general viewpoint, the following molecule classes have to be considered when speaking of *I. paraguariensis* infusions (Heck and Mejia 2007):

(1)   Polyphenols, chlorogenic acid above all
(2)   Xanthines (purine alkaloids)
(3)   Saponins (a group of natural glycosides)
(4)   Minerals: potassium, magnesium, manganese iron, aluminium, and zinc.

With relation to the first class, polyphenols, it has been reported that their amount is strongly dependent on:

(a)   The quality of *yerba mate*
(b)   Particle sizes (the higher the dimension, the lower the superficial interface between grinded leaves and water, and consequently the lower the quantity of extracted molecules)
(c)   And the possible mixture between different *yerba mate* selections.

Consequently, the amount of extractable polyphenols in water is reported to be approximately 92 mg equivalents of chlorogenic acid per each gram of dried leaves, on condition that *yerba mate* belongs to a selection only (no mixed selections). Otherwise, the amount of extractable polyphenols should be reduced (Heck and Mejia 2007). In the ambit of antioxidant capability, *yerba mate* has notable importance because the important polyphenol amount, similar to green tea values (and related performances). However, the analytical determination can give different results depending on the solvent: in fact, organic solvents such as acetone can give significantly higher quantities of extractable polyphenols. The problem is that researchers are generally interested to consumable (and water-extractable) polyphenols, and this amount is a minor fraction of the total quantity of present phenolics (Heck and Mejia 2007; Turkmen et al. 2006). In particular, the most abundant polyphenol in *I. paraguariensis* infusions is chlorogenic acid, while another important class—catechins—is substantially absent (Heck and Mejia 2007).

With reference to xanthines, these substances are common also in tea, chocolate products, and coffee (Heck and Mejia 2007). The most commonly found xanthines in *yerba mate* are caffeine and theobromine above all. Other molecules of interest can be caffeic acid, caffeoylshikimic acid, several caffeoyl derivatives, kaempferol, quercetin, quinic acid, and rutin. The detection of similar phenolics is common to other species in South America and worldwide, including green and black tea leaves. Interestingly, the caffeine amount is very similar to the quantity extractable from coffee: however, 260 or more mg of caffeine per day may be consumed in South America because the normal mate consumption is based on approximately 500 ml of aqueous infusion. On the contrary, coffee cups should be very small. It has to be noted also that processing methods obtaining *yerba mate* can cause the notable decrease of caffeine contents (especially in the *sapeco* and *barbaqua*, or blanching and drying steps), up to 30% if compared with the original quantity. Additionally, chlorophyll decreases strongly with the notable colorimetric turning, as expected. However, and with exclusive relation to caffeine, the repeated infusion and consumption cycle (and the relevant loss of moisture in above-cited steps) can explain the reason which caffeine is still a high amount for, if considered as daily consumption (Heck and Mejia 2007).

With concern to saponins, their importance in vegetable plants such as ginseng (*Panax ginseng*) roots is linked to their ability to disrupt vegetable membranes, being so able to create large micelles with bile acids and steroids. In other words, these compounds can act as surfactant agents. This fact may explain, in part at least, certain features of folk medicines. Chemically, these compounds are subdivided in steroidal and triterpenoid saponins, depending on the aglycone skeleton (Bastos et al. 2007; Heck and Mejia 2007).

Finally, the inorganic (mineral) profile of extractable *yerba mate* remains to be discussed with an introduction (the complete discussion is in Chap. 4). With relation to extractable minerals (iron, aluminium, zinc, and manganese above all), it should be clarified that (Bastos et al. 2007; Heck and Mejia 2007):

(1)   The amount of available minerals depends on agricultural practices and seasons. Consequently, a certain difference/variability between leaves from only-harvested and cultivated/mixed plantations has to be taken into account, partially explaining the limited data amount in the scientific literature

(2)   Secondly, mineral extraction by *yerba mate* roots does not imply the same or similar amount of extracted mineral in leaves. It has been reported that calcium and sodium show opposite behaviours in this ambit, partially because of water solubility.

This situation has to be considered when speaking of variable results concerning *yerba mate* infusions: the temperature and water volumes notably influence the solubility of minerals such as potassium and chlorine, without further factors. Interestingly, it has been reported that the general quantity of extractable minerals in a *mate* infusion increases if tannin amount decreases. Probably, more research is needed in this ambit. On the other hand, it has been reported that *yerba mate* could act as a lead reservoir, but available data should not be worrying, being lower than limits proposed by the United States Environmental Protection Agency (EPA) in 2003 (Heck and Mejia 2007).

1.6   The Safety Importance of *Yerba Mate* and Similar Infusions

Yerba mate beverages have been often reported to have antioxidant power and consequently interesting healthy effects with relation to human safety. Substantially, all above-mentioned bioactive molecules—phenolics, saponins, xanthines—seem to be associated with positive features against several human diseases (Filip et al. 2010; Frizon et al. 2018; Heck and Mejia 2007; Martins et al. 2009). In addition, *I. paraguariensis* contains a broad selection of elements and vitamins, and this fact may have its own importance when speaking of human nutrition and supplementation against nutritional deficiencies.

It should be recognised and highlighted that the regular assumption of *yerba mate* infusions in notable amounts (500 ml at least per day) can explain partially sanitary evidences with reference to antioxidant and hypoglycaemic power. Moreover, cardiovascular diseases linked to atherosclerosis, and the possible anti-cancer treatments should be considered as the effect of the protection of cell membranes and also the genetic code. For these reasons, the research of nutraceutical products has been also concerned the possible addition of *yerba mate* extracts (*chimarrão, tererê*, and *mate* tea) to non-South American typical products or beverages such as soy beverages (Frizon et al. 2018).

On the other side, *yerba mate* has been sometimes reported to be an additive and 'carrier' beverage for characteristic medicinal plants. In detail, it has been reported with reference to the Paraguayans that *I. paraguariensis* is not really considered a medicinal plant. However, its use as infusion is recognised in association with other plants which are widely considered medicinal remedies in Paraguay and in Argentina. Consequently, it is still unknown whether *yerba mate* (as *chimarrão, tererê*, and *mate* tea) has peculiar properties against hypertension, high cholesterol amounts, and diabetes as the result of a pure addition to other medicinal plant extract dissolved in *I. paraguariensis* infusions. The other possibility is that *yerba mate* infusions can show a synergetic function in association with other vegetable remedies. In this ambit, more research is surely needed at present (Arenas and Azorero 1977; Kujawska et al. 2017; Schmeda-Hirschmann and Bordas 1990).

Finally, it has been sometimes reported that the notable *yerba mate* consumption could be linked with increased risk of various tumours (Bates et al. 2007; Heck and Mejia 2007). These findings have certainly confused normal *yerba mate* consumers with obvious results. As a result, there is still a certain contrast between potential health effects of *I. paraguariensis* (as hot or cold infusion) because of phenolics and other bioactive principles on the one side, and the possible carcinogenic effects on the other side (Frizon et al. 2018).

In general with explicit relation to positive claimed effects of *yerba mate* assumption in humans, the following effects have been reported so far (Baeza et al. 2017; Cahuê et al. 2019; Cittadini et al. 2019a, b; Croge et al. 2020; Luís et al. 2019; Panza et al. 2018; Tate et al. 2020):

(a)   Interesting reduction of body weight

(b)   Diminution of oxidative stress
(c)   Cardioprotection evidences
(d)   Glucose reduction and enhancement of sensitivity to insulin
(e)   General antioxidant and antinflammatory features with associated hepatopro-
       tection and neuroprotection
(f)   Reported evidences against retinal degenerations (where oxidative stress plays
       a key role).

These features are dependent on the chemical profile of *I. paraguariensis* infusions
in humans. A more detailed description of these effects and probable causes is offered
in Chap. 5.

On the other side, the possible association between cancer and *yerba mate*
consumption is still unclear, especially when speaking of oral cancer in humans.
Because of contradictory results concerning the possible inhibitory effect against
oral cancer cells in animals and in in vitro studies, it has been suggested that a
possible association exists when tobacco and/or alcohol are included as synergistic
factors with *yerba mate* consumption. At present, there are not reliable assumptions
concerning the augment of cancer in humans depending on the oral consumption
of *I. paraguariensis* infusions (Aguilera 2015; Deneo-Pellegrini et al. 2013; Stefani
et al. 2011).

1.7   *Yerba Mate* Between Regulatory Norms
and Commercial Matters

The economic importance of *yerba mate* harvesting and cultivation in Argentina,
Brazil, Uruguay, and Paraguay has been previously cited. In detail, this plant is a
critical economic resource for South American people working in farms and associated industries. The example of Argentina is notable: *yerba mate* has been defined
the national infusion in this Country in 2013, showing the strict relationship between
I. paraguariensis and this nation. At the same time, other cultivating and harvesting
people in Brazil, Paraguay, and Uruguay have the same or similar feeling (Gortari
2007; Lawson 2009). At present, the real problem with *yerba mate* is not associated with the large commercial operations worldwide: on the contrary, the intensive
relationship between non-South American Nations and commercial production in
South America has generate notable incomes. The problem, by the regulatory angle,
is associated today with the ineffective control and management of the agribusiness
industry, with the result that many small farmers are forced to abandon their homes
and cultivation because of the clear inability to sustain required production volumes.
On the other hand, *yerba mate* market is essentially and historically an oligopolistic
system: the old manners are difficult to be changed. In this ambit, the fair trade
politics and associated movements have partially determined some amelioration for
agricultural workers (Smith 2014). The success of fair trade is dependent from many
factors, including globalisation and the consequent relocation of South American

citizens in other continents such as Europe. In addition, the interest in ethical policies and ethnic products (alternative foods) has determined a strong interest in *yerba mate* products. As a result, fair trade policies have been enhanced in this and other primary production sectors. Anyway, it has to be remembered that the supply chain includes, in the following order: labourers; producers; processors; and final distributors (Ballvé 2007; Boyd et al. 2009). This chain is based on traditional habits, and it is difficult to think about alternative systems at present.

The history of *yerba mate*, from the Precolumbian Age to the Spanish Empire, demonstrated how this plant has obtained favour by new Conquerors: after 1537 A.D. and the birth of Paraguay, the Spanish Crown had managed the harvesting of all existing *yerba mate* in South America, forcing also indigenous *Guaraní* labourers to spend some months per year in this activity (Smith 2014). Harvested native plants would have been used as the local money, or delivered to Spanish provinces such as Potosí (Bolivia). After 1645 A.D., the Jesuits obtained a particular Royal permission for cultivating *yerba mate* in their missions (Paraná-Paraguay area), side-by-side with indigenous labourers. This situation allowed them to export Chile and also Central America (Panama and Mexico). After the expulsion of the Jesuits (1767 A.D.) from South American colonies, the *yerba mate*-based economy remained in the hands of a little oligopoly in Paraguay: the labour force was, as always, the *Guaraní* population. This situation would be notably modified between 1864 and 1870 (the War of the Triple Alliance) with the South American equilibrium shared by Argentina and Brazil. The germination method for *yerba mate* was then discovered again in these two nations, with the result that the modern *yerba mate* production is substantially concentrated here. Curiously, the main area for non-native *yerba mate* in Argentina corresponds to the old Missions of the Jesuits (at present, the Province of *Misiones*) (Smith 2014).

At present, the commercial development and basic rules for *yerba mate* productions are defined, in Argentina, in the ambit of the *Instituto Nacional de la Yerba Mate* (INYM). This Institution rules the whole sector when speaking of commercial types, prices, and the general promotion of *yerba mate* in Argentina and worldwide (Lawson 2009). However, the oligopolistic nature of the *yerba mate* system in South America is still the normal status, meaning substantially that sellers exceed notably the number of buyers. The current situation is the key for the comprehension of the global *yerba mate* market, taking into account the two following variables (Smith 2014):

(1)   *Yerba mate*, in all possible forms, is commercialised as a niche product or—better—as an organic food in Europe (UK, Italy, Germany, Austria, France, and Greece); North America (Canada; USA); Oceania (New Zealand); and Asia (Malaysia, Japan). This list is not exhaustive, but it can give an interesting idea of *yerba mate* sold with an organic certification, allowing for a better diversification of markets and (probably) increased incomes for small farmers. A non-secondary destination, Syria, adsorbs 40% of the total amount of *yerba mate* destined to Middle East countries (Lebanon, Palestine, and Syria), and this situation persists since 1935 at least

(2)   Secondly, this plant can be used as base for flavoured ready-to-drink beverages. The existence of similar products has allowed the *yerba mate* system to find other markets and applications, including also the possible use of *yerba mate* extracts for functional foods

(3)   The institution of electronic commerce has allowed *yerba mate* to arrive rapidly in different world areas where this plant was initially unknown. For these reasons also, *yerba mate* can arrive easily also in the Middle East, probably because of the notable demand from migrants who have know this product while working in South America (Jouanjean et al. 2015).

With relation to commerce, and from the regulatory viewpoint:

(1)   The Argentinean INYM has finally obtained the geographical indication for '*yerba mate*' in 2016 (Moeller IP Advisors 2016). As a consequence, thanks to this recognition, the INYM can now guarantee transparency when speaking of quality distinctions (Sect. 1.4), legal protection, and information related to the origin of *yerba mate* (introducing the concept of traceability and authenticity in this field)

(2)   A 'National Policy on *Yerba mate*' has been issued in Brazil (03 January 2019) with the aim of enhancing the production, quality, and the general commerce for this plant (Presidência da República 2019). The document seems to pursue the same objectives of the Argentinean INYM, including topics such as the defence of environmental sustainability, safety, and hygiene issues with reference to *yerba mate* as a beverage (infusion).

From the international viewpoint, it should be remembered that *yerba mate* can be classified as 'Paraguayan tea', 'botanical herb' or '*mate* tea' according to the Food and Drug Administration (*yerba mate* harmonization code is: 0903.00.00) (Askaripour 2018).

With concern to the European Union (EU), *yerba mate*—under different names— is now recognised as a 'novel food' (European Commission 2020). Originally, *yerba mate* was on the European market as a food or food ingredient, and consequently allowed for a significant consumption until 15 May 1997. At present, the access of this product to the EU market is ruled in accordance with the Novel Food Regulation (EC) No. 258/97. However, it has to be considered that Member States may have peculiar restrictions in this ambit. In accordance with the Trade Part of the Agreement following the agreement in principle announced on 28 June 2019 (Agreement between the EU and the *Mercado Común del Sur* or MERCOSUR) (European Commission 2019), *yerba mate* is considered in the category of 'beverages' when speaking of '*Yerba mate Argentina/ Yerba mate Elaborada con Palo*' (origin: Argentina), 'other beverages' as *Yerbamate Paraguaya*' (Paraguay), or '*mate* herb' with concern to '*São Matheus*' (Brazil).

Anyway, this product has been declared '*Patrimonio Cultural del Mercosur*' in 2018. This declaration shows well the importance of this product for South American consumers. It has to be noted that general features concerning the entering *yerba mate* products on the MERCOSUR market have to comply with MERCOSUR's Technical

Regulation for the Labelling of Packed Food, Resolution No 26/2003. These rules and obligation are discussed again in Chap. 7. Other obligations, such as the Uruguay's *'Decreto No 32/015—Límites de contaminantes en yerba mate'* concerning limits of contaminant in this product, and the main Argentinean regulations ruling *yerba mate* by the INYM should be considered (INYM 2020; Ministerio de Salud Pùblica 2015). The interested Reader is invited to consult the related literature.

1.8   Could *Yerba Mate* Be Associated with Food Frauds?

As above mentioned, the Argentinean INYM has finally obtained the Geographical indication for *'yerba mate'* in 2016 (Moeller IP Advisors 2016). As a consequence, thanks to this recognition, the INYM can now guarantee transparent procedures with reference to commercial typologies, legal recognition, and also origin-related information. In this ambit, the concept of authenticity is a basic pillar, and the same thing can be affirmed when speaking of traceability issues. However, the main question is *'Why should we ask for authenticity and traceability?'* The answer is necessarily related to the possible difference between claimed and declared information concerning a food product and real features of the same food product. In other terms, the nature of this answer is strictly related to food frauds, and it should be considered now whether *yerba mate* may be questioned in some occasions.

The basic concept of intentional adulteration or economically motivated adulteration (Everstine et al. 2013) should imply that one food business operator (FBO) at least in the food supply chain:

(1)   Knows the difference between claimed and real food features
(2)   Is willing to take advantage from the fraudulent exposition of the food product naming it in a non-real way, or claiming one or some specific features this food cannot hold
(3)   Is well aware that the modification and/or omission of certain information can give some economic gain.

Consequently, traceability issues are certainly interesting when speaking of safety risks and correlated analyses at least, including shelf-life information, according to the Parisi's First Law of Food Degradation[1] (FNCF 2020; Parisi 2002a, b, 2003, 2004; Volpe et al. 2015). On the other hand, authenticity concerns are strictly related to food frauds and related gains.

The interest in *yerba mate* adulteration appears quite constant in recent years. The research of terms such as *'yerba mate'* associated with 'adulteration' and 'fraud' on the scholar platform https://scholar.google.com gives 142 results (date: 24 November 2020), including: 7, 10, 13, 10, and 11 references in 2016, 2017, 2018, 2019, and 2020, respectively. There are not particular increasing trends when speaking of these

---

[1]This Law states that 'There are not foods which are not subjected over time to a progressive transformation of their chemical, physical, organoleptic, microbiological, and structural features.

researches; however, the examination of adulteration features seems to show a broad 'choice' of alternative frauds...

By the viewpoint of official controls, some interesting news can be reported. The EU Rapid Alert System for Food and Feed (RASFF) Consumers' Portal, publicly available on the web (https://webgate.ec.europa.eu/rasff-window/consumers/) has mentioned two recent adulteration episodes concerning *yerba mate* in the EU:

(a)   The Notification No. 2019.2199 of 17 June 2019 related to *yerba mate* (considered in the 'cocoa and cocoa preparations, coffee and tea' product category) from Syria, hazard: detection of anthraquinone (unauthorised substance). Web address: https://webgate.ec.europa.eu/rasff-window/portal/?event=notifi cationDetail&NOTIF_REFERENCE=2019.2199

(b)   The Notification No. 2020.0380 of 24 January 2020 related to *yerba mate* (considered in the 'cocoa and cocoa preparations, coffee and tea' product category) from Paraguay, hazard: detection of anthraquinone (unauthorised substance). Web address: https://webgate.ec.europa.eu/rasff-window/portal/?event=notificationDetail&NOTIF_REFERENCE=2020.0380.

These two episodes concern potential hazards from the safety viewpoint. However, it should be considered that:

(1)   The 2019-Notification has concerned Syria as the exporting Country in the EU, while 23 EU countries have received and/or distributed this commodity. In addition, the following non-EU countries are mentioned: Lebanon, Switzerland, Ukraine, and Argentina (the probable origin of *yerba mate*)

(2)   The 2020-Notification has concerned Paraguay as exporting country, and three 'distribution-related' countries in the EU: Germany, Poland (notifying Country), and Slovakia.

These two simple situations have shown that the distribution of *yerba mate* with potential safety problems may be broad enough. Moreover, the 2019-Notification showed apparently *yerba mate* as originated in Syria, while it is evident that this product was only packed in Syria. The real origin was Argentina. Consequently, there is need of accurate and reliable traceability information concerning these products, also because of the possible packaging and labelling in non-South American Countries.

A deep analysis of selected scientific references and above-discussed RASFF Notifications can give more than one single authenticity concern. In detail, the following factors—authenticity menaces—can be highlighted (Crighton et al. 2019; Kucharska-Ambrożej and Karpinska 2020; Lima 2019; Marcelo et al. 2014; Porcari et al. 2016; Poswal et al. 2019; Preti 2019; Santos et al. 2020; Schneider 2017; Sniechowski and Paul 2008; Trentanni Hansen et al. 2019; Vieira et al. 2020):

(1) Undeclared and fraudulent addition of carbohydrates: saccharose, glucose, and fructose. Reason: sensorial enhancement for expired or low-quality *yerba mate* products. Targeted sensorial features: bad appearance; unacceptable or low taste; low weight. Desired results: amelioration of appearance, taste, and augmented weight. Economic gain: reduced raw material amount; saccharose and other simple sugars are cheap enough

(2) Fraudulent identification of origin concerning *yerba mate* products. Reason: economic gain obtained by claiming incorrect Geographical Origin (GI) for *yerba mate*. Targeted sensorial features: none in particular (labelling fraud). Desired results: increase of prices. Economic gain: increase of market prices in spite of the real origin of *yerba mate* products. Note: the fraud may concern not only the mention of a national identity (basically: Argentina, Brazil, Paraguay and Uruguay), but also the mention of a peculiar region or area into the same country

(3) Fraudulent addition and mixing of *I. paraguariensis* with other *Ilex* species. Reasons; economic gain because of the addition of cheap raw materials. Targeted sensorial features: taste (reduced amount of caffeine; different amount of saponins); claimed pharmaceutical properties. Desired results: augment of market amounts for *yerba mate*. Economic gain: the augment of market amounts is balanced with the entering on the market of low-quality and equally-priced *yerba mate*, with resulting low prices for authentic *yerba mate*

(4) Fraudulent labelling in terms of mandatory and facultative information to the consumer.

With relation to the last point, the following data are mandatory in Argentina when speaking of desiccated *yerba mate* or '*yerba*' leaves, with the possible blending with powdered/fragmented dried twigs, floral peduncles, etc. (Sniechowski and Paul 2008):

(a) Name of the food
(b) Brand
(c) Product identification
(d) Lot identification
(e) List of ingredients
(f) Net weight
(g) Best-before date
(h) Data concerning the producer, including the food registration.

In addition, some facultative data may be offered;

(i) Packaging date
(j) Instructions for storage
(k) Instruction for use
(l) Advertising information.

On these bases:

(1)   The identification of the product may vary notably, including the mention of artificial sweeteners, 'simple aromatic herbs', seasoned twigs, flavoured twigs. Other mentions such as 'green', 'digestive anti-acid', 'classic', etc., can be observed

(2)   The mention of non-*yerba mate* ingredients can be omitted unless it does not reach or surpass 25% of the total amount (on the other side, food additives exploiting a technological function have to be declared)

(3)   The use of a pictured country flag on labels may be observed, especially in Brazil. Interestingly, Uruguayan enterprises are accustomed to import raw materials; consequently, the Uruguay flag is not observed or reported at present.

The presence of undeclared or partially mentioned information on labels which is not compliant with above-mentioned requirements in Argentina is a food fraud. The above-mentioned lists can serve as a useful guide in this ambit. The topic is discussed in detail in Chap. 8.

Finally, the following facts have to be highlighted:

(a)   With relation to simple sugars addition, the product '*yerba mate* with saccharose' is allowed in Brazil at least (Ministério da Saúde 2005). Naturally, this quality should be considered as cheap enough if compared with real *yerba mate*

(b)   In addition, the amount of sugars in *yerba mate* has to decrease during processing (blanching). Consequently, the addition of sugars aims at giving back a part of sugars to the original plant, while the normal process operates in the opposite direction with the augment of polyphenols and antioxidants (Dartora et al. 2011)

(c)   Moreover, simple sugars are easily fermentescible, with a theoretical risk in terms of microbial spreading, increased moisture during time, and sanitary damages for diabetics (because of the undeclared sugars presence and addition)

(d)   The entering of low-quality *yerba mate* can destroy the whole market with a general decrease of high-quality plants and leaves. The mass of small and fair traders will necessarily decrease during time

(e)   The addition of artificial colours is not reported so far. Actually, the colorimetric appearance of *yerba mate* should be a good reason for food adulteration. Anyway, there are not evidences for similar practices so far

(f)   Finally, the problem of controls on *yerba mate* is still linked to the lack of accepted analytical methods, on the one hand, and the short shelf-life.

References

Arenas P, Azorero R (1977) Plants of common use in Paraguayan folk medicine for regulating fertility. Econ Bot 31(3):298–300. https://doi.org/10.1007/BF02866879

Aguilera JM (2015) *Ilex paraguariensis* (*Yerba mate*) infusions and risk of oral cancer: a structured literature review. University of Ottawa, Ottawa. Available https://ruor.uottawa.ca/bitstream/10393/33451/1/Yerba%20Mate%20and%20Oral%20Cancer.pdf. Accessed 10 Dec 2020

Askaripour D (2018) How to travel with *yerba mate*—safely and legally. Circle of drink, Inc., New York. Available https://circleofdrink.com/how-to-travel-with-yerba-mate. Accessed 11 Dec 2020

Ballvé T (2007) Mate on the market: fair trade and the gaucho's 'liquid vegetable'. NACLA Rep Am 40(5):10–13. https://doi.org/10.1080/10714839.2007.11722293

Baeza G, Sarriá B, Bravo L, Mateos R (2017) Polyphenol content, in vitro bioaccessibility and antioxidant capacity of widely consumed beverages. J Sci Food Agric 98(4):1397–1406. https://doi.org/10.1002/jsfa.8607

Bastos DHM, De Oliveira DM, Matsumoto RT, Carvalho PDO, Ribeiro ML (2007) *Yerba mate*: pharmacological properties, research and biotechnology. Med Aromat Plant Sci Biotechnol 1(1):37–46

Bates MN, Hopenhayn ROA, Moore LE (2007) Bladder cancer and mate consumption in Argentina: a case-control study. Cancer Lett 246(1–2):268–273. https://doi.org/10.1016/j.canlet.2006.03.005

Bhagat AR, Delgado AM, Issaoui M, Chammem N, Fiorino M, Pellerito A, Natalello S (2019) Review of the role of fluid dairy in delivery of polyphenolic compounds in the diet: chocolate milk, coffee beverages, matcha green tea, and beyond. J AOAC Int 102(5):1365–1372. https://doi.org/10.1093/jaoac/102.5.1365

Boyd B, Henning N, Reyna E, Wang D, Welch M (2009) Hybrid organizations: new business models for environmental leader-ship. Greenleaf Publishing, Sheffield

Bracesco N, Sanchez AG, Contreras V, Menini T, Gugliucci A (2011) Recent advances on Ilex paraguariensis research: minireview. J Ethnopharmacol 136(3):378–384. https://doi.org/10.1016/j.jep.2010.06.032

Barbieri G, Bergamaschi M, Saccani G, Caruso G, Santangelo A, Tulumello T, Vibhute B, Barbieri G (2019) Processed meat and polyphenols: opportunities, advantages, and difficulties. J AOAC Int 102(5):1401–1406. https://doi.org/10.1093/jaoac/102.5.1401

Barbera M (2020) Reuse of food waste and wastewater as a source of polyphenolic compounds to use as food additives. J AOAC Int 103(4):906–914. https://doi.org/10.1093/jaocint/qsz025

Cansiani R, Mossi A, Mosele S, Toniazzo G, Treichel H, Paroul N, Oliveira JV, Oliveira D, Mazutti M, Echeverrigaray S (2008) Genetic conservation and medicinal properties of mate (Ilex paraguariensis St Hil.). Phcog Rev 2(4):326–328

Chaimsohn FP, Machado NC, Gomes EP, Vogt GA, Neppel G, Souza AM, Marques AC (2014) *Yerba mate*'s traditional and agroforestry systems and impacts on territorial development: the center-south of Paraná and northern Santa Catarina. In: Dallabrida VR (ed) Territorial development: Brazilian public policies, international experiences and geographical indication as a reference. LiberArs, São Paulo, pp 47–54

Chammem N, Issaoui M, De Almeida AID, Delgado AM (2018) Food crises and food safety incidents in European Union, United States, and Maghreb area: current risk communication strategies and new approaches. J AOAC Int 101(4):923–938. https://doi.org/10.5740/jaoacint.17-0446

Cittadini MC, Albrecht C, Miranda AR, Mazzuduli GM, Soria EA, Repossi G (2019a) Neuroprotective effect of Ilex paraguariensis intake on brain myelin of lung adenocarcinoma-bearing male Balb/c mice. Nutr Cancer 71(4):629–633. https://doi.org/10.1080/01635581.2018.1559932

Cahuê F, Nascimento JHM, Barcellos L, Salerno VP (2019) Ilex paraguariensis, exercise and cardio-protection: a retrospective analysis. J Funct Foods 53:105–108. https://doi.org/10.1016/j.jff.2018.12.008

Cittadini MC, Repossi G, Albrecht C, Di Paola NR, Miranda AR, Pascual-Teresa S, Soria EA (2019b) Effects of bioavailable phenolic com-pounds from Ilex paraguariensis on the brain of mice with lung adenocarcinoma. Phytother Res 33(4):1142–1149. https://doi.org/10.1002/ptr.6308

Clarence-Smith WG, Topik S (eds) (2003) The global coffee economy in Africa, Asia, and Latin America, 1500–1989. Cambridge University Press, Cambridge

Crighton E, Coghlan ML, Farrington R, Hoban CL, Power MW, Nash C, Mullaney I, Byard RW, Trengove R, Musgrave IF, Bunce M, Maker G (2019) Toxicological screening and DNA sequencing detects contamination and adulteration in regulated herbal medicines and supplements for diet, weight loss and cardiovascular health. J Pharm Biomed Anal 176:112834. https://doi.org/10.1016/j.jpba.2019.112834

Croge CP, Cuquel FL, Pintro PTM (2020) Yerba mate: cultivation systems, processing and chemical composition. A Rev Sci Agricola 78(5):e20190259. https://doi.org/10.1590/1678-992X-2019-0259

da Silveira TFF, de Souza TCL, Carvalho AV, Ribeiro AB, Kuhnle GGC, Godoy HT(2017) White açaí juice (Euterpe oleracea): phenolic composition by LC-ESI-MS/MS, antioxidant capacity and inhibition effect on the formation of colorectal cancer related compounds. J Funct Foods 36:215–223. https://doi.org/10.1016/j.jff.2017.07.001

Dartora N, de Souza LM, Santana AP, Iacomini M, Valduga AT, Gorin PAJ, Sassaki GL (2011) UPLC-PDA–MS evaluation of bioactive compounds from leaves of Ilex paraguariensis with different growth conditions, treatments and ageing. Food Chem 129(4):1453–1461. https://doi.org/10.1016/j.foodchem.2011.05.112

de Camargo AC, Schwember AR (2019) Phenolic-driven sensory changes in functional foods. J Food Bioact 5:6–7. https://doi.org/10.31665/JFB.2019.5173

De Toffoli A, Monteleone E, Bucalossi G, Veneziani G, Fia G, Servili M, Zanoni B, Pagliarini E, Gallina Toschi T, Dinnella C (2019) Sensory and chemical profile of a phenolic extract from olive mill waste waters in plant-base food with var-ied macro-composition. Food Res Int 119:236–243. https://doi.org/10.1016/j.foodres.2019.02.005

Delacassa E, Bandoni AL (2001) El mate. Revista de Fitoterapia 1(4):257–265

Delgado AM, Vaz de Almeida MD, Parisi S (2016) Chemistry of the mediterranean diet. Springer Int Publishing, Cham https://doi.org/10.1007/978-3-319-29370-7

Delgado AM, Almeida MDV, Parisi S (2017) Chemistry of the mediterranean diet. Springer Int Publishing, Cham https://doi.org/10.1007/978-3-319-29370-7

Delgado AM, Issaoui M, Chammem N (2019) Analysis of main and healthy phenolic compounds in foods. J AOAC Int 102(5):1356–1364. https://doi.org/10.1093/jaoac/102.5.1356

Deneo-Pellegrini H, De Stefani E, Boffetta P, Ronco AL, Acosta G, Correa P, Mendilaharsu M (2013) Maté consumption and risk of oral cancer: case-control study in Uruguay. Head Neck 35(8):1091–1095. https://doi.org/10.1002/hed.23080

Dibanda RF, Akdowa EP, Tongwa QM (2020) Effect of microwave blanching on antioxidant activity, phenolic compounds and browning behaviour of some fruit peelings. FoodChem 302:125308. https://doi.org/10.1016/j.foodchem.2019.125308

Efing LC, Caliari TK, Nakashima T, De Freitas RJS (2009) Caracterização química e capacidade antioxidante da erva-mate (Ilex paraguariensis st. Hil.). Boletim do Centro de pesquisa de Processamento de Alimentos 27(2):241–246. https://doi.org/10.5380/cep.v27i2.22034

European Commission (2019) New EU-mercosur trade agreement—the agreement in principle, Brussels, 1 July 2019. European Commission, Brussels. Available https://trade.ec.europa.eu/doclib/docs/2019/june/tradoc_157964.pdf. Accessed 11 Dec 2020

European Commission (2020) EU novel food catalogue. European commission, Brussels. Available https://ec.europa.eu/food/safety/novel_food/catalogue/search/public. Accessed 11 Dec 2020

Everstine K, Spink J, Kennedy S (2013) Economically motivated adulteration (EMA) of food: common characteristics of EMA incidents. J Food Prot 76(4):723–725. https://doi.org/10.4315/0362-028X.JFP-12-399

Filip R, Davicino R, Anesini C (2010) Antifungal activity of the aqueous extract of Ilex paraguariensis against Malassezia furfur. Phytother Res 24(5):715–719. https://doi.org/10.1002/ptr.3004

Fiorino M, Barone C, Barone M, Mason M, Bhagat A (2019) The intentional adulteration in foods and quality management systems: chemical aspects. In: Quality systems in the food industry. Springer International Publishing, Cham pp 29–37

Flamminii F, Di Mattia CD, Difonzo G, Neri L, Faieta M, Caponio F, Pittia P (2019) From by-product to food ingredient: evaluation of compositional and technological properties of olive-leaf phenolic extracts. J Sci Food Agric 99(14):6620–6627. https://doi.org/10.1002/jsfa.9949

FNCF (2020) La Prima Legge della degradazione alimenti prende il nome da un chimico italiano. Federazione Nazionale dei chimici e dei fisici (FNCF), Roma. Available https://www.chimic ifisici.it/la-prima-legge-della-degradazione-alimenti-prende-il-nome-da-un-chimico-italiano/. Accessed 11 Dec 2020

Folch C (2010) Stimulating consumption: *yerba mate* myths, markets, and meanings from conquest to present. Comp Stud Soc Hist 52(1):6–36. https://doi.org/10.1017/S0010417509990314

Frizon CNT, Perussello CA, Sturion JA, Hoffmann-Ribani R (2018) Novel beverages of yerba-mate and soy: bioactive compounds and functional properties. Beverages 4(1):21. https://doi.org/10.3390/beverages4010021

Garcia-Lazaro RS, Lamdan H, Caligiuri LG, Lorenzo N, Berengeno AL, Ortega HH, Alonso DF, Farina HG (2020) In vitro and in vivo antitumor activity of *Yerba mate* extract in colon cancer models. J Food Sci 85(7):2186–2197. https://doi.org/10.1111/1750-3841.15169

Giberti GC (1994) Mate (Ilex paraguariensis). In: Hernando BJE, León J (eds) Neglected crops: 1492 from a different perspective. Plant production and protection series no. 26. Food and Agriculture Organization of the United Nations (FAO), Rome, Italy pp 252–254

Giberti GC (1995) Aspectos oscuros de la corología de Ilex paraguariensis St. Hil. In: Winge H, Ferreira AG, de Araujo Mariath JE, Tarasconi LC (eds) "Erva-Mate: biologia e cultura no cone sul. Editora da UFRGS, Porto Alegre pp 289–300

Goeddel C (2015) *Yerba mate* in Argentina: a cultural reflection and projection through a popular pastime. Proceedings of the 2015 symposium on undergraduate research and creative expression, 14 April 2015, Iowa state university, Ames, IA

Goldenberg D, Golz A, Joachims HZ (2003) The beverage maté: a risk factor for cancer of the head and neck. Head Neck 25(7):595–601. https://doi.org/10.1002/hed.10288

Gortari J (2007) El instituto nacional de la *yerba mate* (INYM) como dispositivo político de economía social". In: Gortari J (ed) De la tierra sin mal al tractorazo. Hacia una economía política de la *yerba mate*. Ed. Universitaria, Universidad Nacional de Misiones, Posadas

Goyke N (2017) Traditional medicinal use in Chamorro Cué, Gral. E. Aquino, San Pedro, Paraguay. Dissertation, Michigan Technological University, Houghton

Greizerstein EJ, Giberti GC, Poggio L (2004) Cytogenetic studies of Southern South-American Ilex. Caryologia 57(1):19–23. https://doi.org/10.1080/00087114.2004.10589367

Grigioni G, Carduza F, Irurueta M, Pensel N (2004) Flavour characteristics of Ilex paraguariensis infusion, a typical Argentine product, assessed by sensory evaluation and electronic nose. J Sci Food Agric 84:427–432. https://doi.org/10.1002/jsfa.1670

Habtemariam S (2019) The chemical and pharmacological basis of yerba maté (Ilex paraguariensis A. St.-Hil.) as potential therapy for type 2 diabetes and meta-bolic syndrome. In: Habtemariam S (ed) Medicinal foods as potential therapies for type-2 diabetes and associated diseases. Academic Press, New York, NY pp 943–983

Haddad MA, Parisi S (2020) The next big HITS. New Food Mag 23(2):4

Haddad MA, Parisi S (2020) Evolutive profiles of mozzarella and vegan cheese during shelf-life. Dairy Ind Int 85(3):36–38

Haddad MA, El-Qudah J, Abu-Romman S, Obeidat M, Iommi C, Jaradat DSM (2020) Phenolics in mediterranean and middle east important fruits. J AOAC Int 103(4):930–934. https://doi.org/10.1093/jaocint/qsz027

Haddad MA, Dmour H, Al-Khazaleh JFM, Obeidat M, Al-Abbadi A, Al-Shadaideh AN, Almazra'awi MS, Shatnawi MA, Iommi C (2020) Herbs and medicinal plants in Jordan. J AOAC Int 103(4):925–929. https://doi.org/10.1093/jaocint/qsz026

Hanisch AL, Vogt GA, Marques AC, Bona LC, Bosse DD (2010) Structure and floristic composition of five caiva area in north plateau of Santa Catarina State, Brazil. Pesquisa Flore-Stal Bras 30:303–310. https://doi.org/10.4336/2010.pfb.30.64.303

Heck CI, De Mejia EG (2007) Yerba mate Tea (Ilex paraguariensis): a comprehensive review on chemistry, health implications, and technological considerations. J Food Sci 72(9):R138–R151. https://doi.org/10.1111/j.1750-3841.2007.00535.x

Ilany T, Ashton MS, Montagnini F, Martinez C (2010) Using agroforestry to improve soil fertility: effects of intercropping on Ilex paraguariensis (yerba mate) plantations with Araucaria angustifolia. Agrofor Syst 80(3):399–409. https://doi.org/10.1007/s10457-010-9317-8

INYM (2020) Normativa vigente. Instituto nacional de la yerba mate (INYM), Posadas. Available https://inym.org.ar/normativa-vigente.html. Accessed 11 Dec 2020

Issaoui M, Delgado AM, Caruso G, Micali M, Barbera M, Atrous H, Ouslati A, Chammem N (2020) Phenols, flavors, and the mediterranean diet. J AOAC Int 103(4):915–992. https://doi.org/10.1093/jaocint/qsz018

Janda K, Jakubczyk K, Łukomska A, Baranowska-Bosiacka I, Rębacz-Maron E, Dec K, Kochman J, Gutowska I (2020) Effect of the Yerba mate (Ilex paraguariensis) brewing method on the content of selected elements and antioxidant potential of infusions. Pol J Chem Technol 22(1):54–60. https://doi.org/10.2478/pjct-2020-0008

Jouanjean MA, Mendez-Parra M, Te Velde DW (2015) Trade policy and economic transformation. Supporting economic transformation (SET) brief. Overseas development institute, London. Available https://set.odi.org/wp-content/uploads/2015/08/Trade-and-economic-transformation-13-July-2015.pdf. Accessed 11 Dec 2020

Karak T, Bhagat RM (2010) Trace elements in tea leaves, made tea and tea infusion: A review. Food Res Int 43(9):2234–2252. https://doi.org/10.1016/j.foodres.2010.08.010

Kucharska-Ambrożej K, Karpinska J (2020) The application of spectroscopic techniques in combination with chemometrics for detection adulteration of some herbs and spices. Microchem J 153:104278. https://doi.org/10.1016/j.microc.2019.104278

Kujawska M (2018) Yerba mate (Ilex paraguariensis) beverage: nutraceutical ingredient or conveyor for the intake of medicinal plants? Evidence from Paraguayan folk medicine. Evid Based Compl Alt 2018:6849317. https://doi.org/10.1155/2018/6849317

Kujawska M, Hilgert NI (2014) Phytotherapy of polish migrants in misiones, Argentina: legacy and acquired plant species. J Ethnopharmacol 153(3):810–830. https://doi.org/10.1016/j.jep.2014.03.044

Kujawska M, Jiménez-Escobar MND, Nolan JM, Arias-Mutis D (2017) Cognition, culture and utility: plant classification by paraguayan immigrant farmers in misiones, Argentina. J Ethnobiol Ethnomed 13(1):42. https://doi.org/10.1186/s13002-017-0169-4

Lawson J (2009) Cultivating green gold: a political ecology of land use changes for small yerba mate farmers in misiones, Argentina. Dissertation, Yale School of Forestry, New Haven, CT

Lima PCD (2019) Discriminação de erva-mate para chimarrão quanto à origem geográfica e presença de açúcar utilizando FTIR e quimiometria. Dissertation, Universidade Tecnológica Federal do Paraná, Curitiba

López A (1974) The economics of yerba mate in seventeenth-century South America. Agric History 48(4):493–509

Lovera NN, Alegre CA, Hedman JC, Surkan SA, Schmalko ME (2019) Determination and intake temperature of the beverage during hot mate consumption. Lat Am Appl Res Int J 49(1):41–45

Luís AFS, Domingues FC, Pereira Amaral LMJ (2019) The anti-obesity potential of Ilex paraguariensis: results from a meta-analysis. Braz J Pharm Sci 55:e17615. https://doi.org/10.1590/s2175-97902019000217615

Maccari Junior A (2005) Análise do pré-processamento da erva-mate para chimarrão. Universidade Estadual de Campinas, Campinas

Magri E, Valduga AT, Gonçalves IL, Barbosa JZ, de Oliveira Rabel D, Menezes IMNR, de Andrade Nascimento P, Oliveira A, Studart Corrêa R, Motta ACV (2020) Cadmium and lead concentrations and *yerba mate* leaves from agroforestry and plantation systems: an international survey in South America. J Food Comp Anal 103702. https://doi.org/10.1016/j.jfca.2020.103702

Marcelo MCA, Martins CA, Pozebon D, Ferrão MF (2014) Methods of multivariate analysis of NIR reflectance spectra for classification of *yerba mate*. Anal Methods 6(19):7621–7627. https://doi.org/10.1039/C4AY01350F

Marques AC, Mattos AG, Bona LC, Reis MS (2012) National forests and research development: *yerba mate* management (*Ilex paraguariensis* A. St.-Hil.) in flona in Tres Barras/Santa Catarina. Biodiversidade Brasileira 2:4–17

Martins F, Suzan AJ, Cerutti SM, Arçari DP, Ribeiro ML, Bastos DH, Carvalho Pde O (2009) Consumption of mate tea (Ilex paraguariensis) decreases the oxidation of unsaturated fatty acids in mouse liver. Br J Nutr 101(4):527–532. https://doi.org/10.1017/S000711450802504X

Matta FV (2019) Chemical analysis of typical beverages and açaí berry from South America dissertation. University of Surrey, Guildford

Ministério da Saúde (2005) RDC no. 277 de 22 de setembro de 2005. Diário Oficial da União (DOU) of 23 Sep 2005 No 184, sexta-feira pp 379–380

Ministero de Salud Pùblica (2015) Decreto No 32/015—límites de contaminantes en *yerba mate*. Ministero de salud pùblica, Montevideo, Uruguay. Available https://www.fao.org/faolex/results/details/en/c/LEX-FAOC143804/. Accessed 11 Dec 2020

Moeller IP Advisors (2016) Argentina: *yerba mate* is a new GI in Argentina. Mondaq® Ltd, London, New York, and Sidney. Available https://www.mondaq.com/argentina/trademark/552298/yerba-mate-is. Accessed 11 Dec 2020

Oberti F (1960) Disquisiciones sobre el origen de la bombilla. Cuadernos Del Instituto Nacional De Antropología Y Pensamiento Latinoamericano 1:151–158

Olivari I, Paz S, Gutiérrez ÁJ, González-Weller D, Hardisson A, Sagratini G, Rubio C (2020) Macroelement, trace element, and toxic metal levels in leaves and infusions of *yerba mate* (*Ilex paraguariensis*). Environ Sci Pollut Res 27:21341–21352. https://doi.org/10.1007/s11356-020-08681-9

Pagliosa CM, Pereira SM, Vieira MA, Costa LA, Teixeira E, Amboni RD, Amante ER (2009) Bitterness in yerba maté (Ilex paraguariensis) leaves. J Sens Stud 24:415–426. https://doi.org/10.1111/j.1745-459X.2009.00218.x

Panza VP, Brunetta HS, Oliveira MV, Nunes EA, Silva EL (2018) Effect of mate tea (Ilex paraguariensis) on the expression of the leukocyte NADPH oxidase subunit p47[phox] and on circulating inflammatory cytokines in healthy men: a pilot study. Int J Food Sci Nutr 70:212–221. https://doi.org/10.1080/09637486.2018.1486393

Panzella L, Moccia F, Nasti R, Marzorati S, Verotta L, Napolitano A (2020) Bioactive phenolic compounds from agri-food wastes: an update on green and sustainable extraction methodologies. Front Nutr 7:60. https://doi.org/10.3389/fnut.2020.00060

Parisi S (2002a) I fondamenti del calcolo della data di scadenza degli alimenti: principi ed applicazioni. Ind Aliment 41(417):905–919

Parisi S (2002b) Profili evolutivi dei contenuti batterici e chimico-fisici in prodotti lattiero-caseari. Ind Aliment 41(412):295–306

Parisi S (2003) Evoluzione chimico-fisica e microbiologica nella conserva-zione di prodotti lattiero-caseari. Ind Aliment 42(423):249–259

Parisi S (2004) The prediction of shelf life about cheese on the basis of storage temperature. Italian J food Sci Special Issue 11–19

Parisi S (2012) Food packaging and food alterations. The User-Oriented Approach, Smithers Rapra Technology Ltd, Shawbury, Shropshire

Parisi S (2013) Food industry and packaging materials—performance-oriented guidelines for users. Smithers Rapra Technologies, Shawbury

Parisi S (2019) Analysis of major phenolic compounds in foods and their health effects. AOAC J 102(5):1354–1355. https://doi.org/10.5740/jaoacint.19-0127

Parisi S (2020) Characterization of major phenolic compounds in selected foods by the technological and health promotion viewpoints. J AOAC Int 103(4):904–905. https://doi.org/10.1093/jaoacint/qsaa011

Parisi S, Dongo D (2020) Polifenoli e salute. I vegetali amici del sistema immunitario. Great Italian food trade. Available https://www.greatitalianfoodtrade.it/salute/polifenoli-e-salute-i-veg etali-amici-del-sistema-immunitario. Accessed 9 Dec 2020

Parisi S, Dongo D, Parisi C (2020) Resveratrolo, conoscenze attuali e prospettive. Great Italian food trade 27/10/2020. Available www.greatitalianfoodtrade.it/integratori/resveratrolo-conoscenze-att uali-e-prospettive. Accessed 9 Dec 2020

Piccolo GA, Galantini JA, Rosell RA (2004) Organic carbon fractions in a *yerba mate* plantation on a subtropical Kandihumult of Argentina. Geoderma 123(3–4):333–341. https://doi.org/10.1016/j.geoderma.2004.02.017

Porcari AM, Fernandes GD, Barrera-Arellano D, Eberlin MN, Alberici RM (2016) Food quality and authenticity screening via easy ambient sonic-spray ionization mass spectrometry. Anal 141(4):1172–1184. https://doi.org/10.1039/C5AN01415H

Poswal FS, Russell G, Mackonochie M, MacLennan E, Adukwu EC, Rolfe V (2019) Herbal teas and their health benefits: a scoping review. Plant Foods Hum Nutr 74:266–276. https://doi.org/10.1007/s11130-019-00750-w

Presidência da República (2019) Lei n. 13.791—dispõe sobre a política nacional da erva-mate. Presidência da república, casa civil, subchefia para assuntos jurídicos, Brazil. Diário oficial da União (DOU) of 4 Jan 2019. Available https://www.fao.org/faolex/results/details/en/c/LEX-FAOC183890/. Accessed 11 Dec 2020

Prestes SLC, Thewes FR, Sautter CK, Brackmann A (2014) Storage of yerba maté in controlled atmosphere. Ciência Rural 44(4):740–745. https://doi.org/10.1590/S0103-84782014000400028

Preti R (2019) Progress in beverages authentication by the application of analytical techniques and chemometrics. In: Grumezescu AM, Holban AM (eds) Quality control in the beverage industry, vol 17: the science of beverages. Academic Press. Sawston. pp 85–121 https://doi.org/10.1016/B978-0-12-816681-9.00003-5

Santa Cruz MJ, Garitta L, Hough G (2002) Sensory de-scriptive analysis of yerba maté (Ilex paraguariensis Saint Hilaire), a South American beverage. Food Sci Technol Int 8(1):25–31. https://doi.org/10.1106/108201302022942

Santos MCD, Azcarate SM, Lima KMG, Goicoechea HC (2020) Fluorescence spectroscopy application for Argentinean *yerba mate* (*Ilex paraguariensis*) classification assessing first-and second-order data structure properties. Microchem J 155:104783. https://doi.org/10.1016/j.microc.2020.104783

Schmeda-Hirschmann G, Bordas E (1990) Paraguayan medicinal compositae. J Ethnopharmacolvol 28(2):163–171. https://doi.org/10.1016/0378-8741(90)90026-P

Schneider M (2017) Determinação da adulteração da erva-mate por adição de sacarose empregando espectroscopia no infravermelho (atr-ftir) em conjunto com ferramentas quimiométricas. Dissertation, Universidade Federal do Rio Grande do Sul, Porto Alegre

Shahidi F, Varatharajan V, Oh WY, Peng H (2019) Phenolic compounds in agri-food by-products, their bioavailability and health effects. J Food Bioact 5(1):57–119. https://doi.org/10.31665/JFB.2019.5178

Singla RK, Dubey AK, Garg A, Sharma RK, Fiorino M, Ameen SM, Haddad MA, Al-Hiary M (2019) Natural polyphenols: chemical classification, definition of classes, subcategories, and structures. J AOAC Int 102(5):1397–1400. https://doi.org/10.1093/jaoac/102.5.1397

Sloan E, Adam Hutt C (2012) Beverage trends in 2012 and beyond. Agro FOOD Industry Hi Tech 23(4):8–12

Small E, Catling PM (2001) Blossoming treasures of biodiversity: 3. Mate (Ilex paraguariensis)—better than Viagra, marijuana, and coffee? Biodivers 2:26–27

Smith F (2014) Exploring fair trade *yerba mate* networks in misiones, Argentina. Dissertation, University of Miami, Miami. Available https://scholarship.miami.edu/discovery/fulldisplay/alm

a991031447938602976/01UOML_INST:ResearchRepository?tags=scholar. Accessed 10 Dec 2020

Sniechowski VI, Paul LM (2008) The labeling on the "*yerba mate*" (*Ilex paraguariensis*) packages in the Mercosur" (South American common market). Visión De Futuro 9(1):105–124

Srivastava PK (2019) Status report on bee keeping and honey processing. Status report on bee keeping and honey processing. Development institute, ministry of micro, small & medium enterprises (MSME), Government of India 107, Industrial Estate, Kalpi road, Kanpur-208012. Available https://msmedikanpur.gov.in/cmdatahien/reports/diffIndustries/Status%20Report%20on%20Bee%20keeping%20&%20Honey%20Processing%202019-2020.pdf. Accessed 9 Dec 2020

Stefani ED, Moore M, Aune D, Deneo-Pellegrini H, Ronco AL, Boffetta P, Correa P, Acosta G, Mendilaharsu M, Luaces ME, Silva C, Landó G (2011) Maté consumption andrisk of cancer: a multi-site case-control study in Uruguay. Asian Pac J Cancer Prev 12(4):1089–1093

Tate PS, Marazita MC, Marquioni-Ramella MD, Suburo AM (2020) Ilex paraguariensis extracts and its polyphenols prevent oxidative damage and senescence of human retinal pigment epithelium cells. J Funct Food 67:103833. https://doi.org/10.1016/j.jff.2020.103833

Trentanni Hansen GJ, Almonacid J, Albertengo L, Rodriguez MS, Di Anibal C, Delrieux C (2019) NIR-based sudan I to IV and para-red food adulterants screening. Food Addit Contam Part A 36(8):1163–1172. https://doi.org/10.1080/19440049.2019.1619940

Turkmen N, Sari F, Velioglu YS (2006) Effects of extraction solvents on concentration and antioxidant activity of black and black Mate tea polyphenols determined by ferrous tartrate and folin-ciocalteu methods. Food Chem 99:835–841. https://doi.org/10.1016/j.foodchem.2005.08.034

Vieira TF, Makimori GYF, dos Santos Scholz MB, Zielinski AAF, Bona E (2020) Chemometric approach using ComDim and PLS-DA for discrimination and classification of commercial *yerba mate* (*Ilex paraguariensis* St. Hil.). Food Anal Methods 13(1):97–107. https://doi.org/10.1007/s12161-019-01520-9

Vogt GA, Neppel G, Souza AMA (2016) The mate activity in north plateau, Santa Catarina state: the geographical indication as an alternative to valorization of *yerba mate*. Desenvolvimento Reg Em Debate 6:64–87

Volpe MG, Di Stasio M, Paolucci M, Moccia S (2015) Polymers for food shelf-life extension. Functional polymers in food science, Scrivener Publishing LLC. pp 9–66

Wight W (1962) Tea classification revised. Curr Sci 31:295–299

Yamaguchi KKDL, Pereira LFR, Lamarâo CV, Lima ES, da Veiga-Junior VF (2015) Amazon acai: chemistry and biological activities: a review. Food Chem 179:137–151. https://doi.org/10.1016/j.foodchem.2015.01.055

Chapter 2
Botanic Features of *Ilex paraguariensis*

Abstract The classification of *Ilex paraguariensis*, generally named *yerba mate* from the botanical viewpoint, is of critical importance because of the need to distinguish this plant from other vegetables of the same family. The botanical subdivision is helpful because of three main factors: the quality of *yerba mate*; commercial features of available *yerba mate* products; and the possible addition of *Ilex*-related non-*yerba mate* plants as the signal of food adulteration. *I. paraguariensis* belongs to the Aquifoliaceae family, and its natural habitat comprehends regional area of four South American countries: Brazil, Argentina, Paraguay, and Uruguay. This four-country area is peculiar with concern to *yerba mate* because this plant is a subtropical dioecious evergreen tree, normally growing in mountainous areas and requiring at least 1200 mm of yearly rainfall. From the botanical viewpoint, this plant needs high aluminium contents, acidic soils, low phosphorus, and abundance of bioavailable organic materials. The taxonomic classification of *Ilex* species may be correlated with chemometrics and fingerprinting methods, demonstrating the interest in the preservation of preserving quality and authenticity of real *yerba mate*. The historical confusion between *I. paraguariensis* and *I. argentina* (a native Bolivian plant) can be a useful example.

Keywords Aquifoliaceae · Fingerprinting · *Ilex argentina* · *Ilex paraguariensis* · Plant · South America · Yerba mate

2.1 The Botanical Classification of *Yerba Mate*: A General Introduction

The classification of *Ilex paraguariensis*, generally named *yerba mate* or *mate* (Botanical-Online 2019; Kim et al. 2010) from the botanical viewpoint is of critical importance because of the need to distinguish this plant from other vegetables of the same family. The botanical subdivision is helpful because of three main factors:

(a) The quality of *yerba mate* is strongly dependent on the original plant or mixtures of different *I. paraguariensis* cultivated plants. Because of the possible mixture option, it has to be noted that the distinction between authentic

© The Author(s), under exclusive license to Springer Nature Switzerland AG 2021   35
C. Iommi, *Chemistry and Safety of South American Yerba Mate Teas*,
Chemistry of Foods, https://doi.org/10.1007/978-3-030-69614-6_2

yerba mate and *Ilex*-related non-*yerba mate* plants is critical when speaking of quality (of the final product).

(b)   The quality of commercially available *yerba mate* products can vary enough because of the existence of eight types: *premium; nativa; traditional; moida grossa; exportação; tererê; tostada; and cambona* (Sect. 1.4.1). These typologies depend on several factors: processing options, national or regional origin, possible genetic variations (the *cambona* type). The possible addition of *Ilex*-related non-*yerba mate* plants could complicate the situation.

(c)   Finally, the addition of *Ilex*-related non-*yerba mate* plants is a clear food adulteration.

Ilex paraguariensis belongs to the *Aquifoliaceae* (holy plants, more than 600 species) family. Its natural *habitat* comprehends the following South American countries (Fig. 1.3): Brazil, Argentina, Paraguay, and Uruguay, taking into account that only some areas are interested in this ambit (Uruguay is completely interested). In detail, the regional areas of interest are known as follows, with the exception of Uruguay (Heck and De Mejia 2007; Matta 2019):

(a)   Brazil: *Mato Grosso do Sul; Minas Gerais; Paraná; Rio Grande do Sul; Rio de Janeiro; Santa Catarina; Sao Paulo*
(b)   Argentina: *Corrientes; Misiones*
(c)   Paraguay: *Alto Parana; Amambay; Caaguazu; Canendeyu; Central; Guaira; Itapua; Misiones; San Pedro.*

More exactly, the reported boundaries of this four-state area are (Croge et al. 2020):

(a)   Latitudes, 21°00′00″ S, 30°00′00″ S
(b)   Longitudes, 48°30′00″ W, 56°10′00″ W.

The natural habitat of *I. paraguariensis* depends on its nature: this plant is a subtropical dioecious evergreen tree (with male and female trees, being the male plants more productive than female ones) (Botanical-Online 2019; Kim et al. 2010), normally growing in mountainous areas. As a result, it is not naturally found in the northern or southern areas of South America for climatic reasons, requiring at least 1,200 mm of yearly rainfall, possibly with a constant distribution during the year. In addition, it can grow well at 21–22 °C, but the resistance to extreme temperatures such as—6 °C and possible snowfalls, is a distinctive advantage (Heck and De Mejia 2007; Matta 2019).

From the botanical viewpoint, it should be noted that *I. paraguariensis* can arrive to 15 m, with peculiar leaves (bright-green colour, alternate or oval shape, length: 5–18 cm; width: 7.2 cm). It flourishes from October to November, and related fruits (dark-to-red drupes, 4–6 mm) are produced in the March–June period.

It can be also noted that South American *Ilex* species and *I. paraguariensis,* in particular, are endemic and exclusively distributed in wild forested mountains, as an association of ombrophilous plants, in association with *Araucaria angustifolia*. These plants need also high aluminium contents, acidic soils, low phosphorus, and abundance of bioavailable organic materials (Croge et al. 2020).

2.2 The Problem of *Ilex*-Related Non-Yerba Mate Plants, Between Confusion and Adulterations

Other *Ilex* genus plants are available in South America and also in East Asia. In relation to discrimination between different *Ilex* types found in South America only, the genus includes the following types which could be of interest because of the possible confusion with *I. paraguariensis var. paraguariensis* St. Hill (Choi et al. 2005):

- *I. argentina* Lillo
- *I. brasiliensis* (Spreng) Loes.
- *I. brevicuspis* Reissek
- *I. dumosa* var. *dumosa* Reissek
- *I. dumosa* var. *guaranina* Loes.
- *I. integerrima* (Vellozo) Reissek
- *I. microdonta* Reissek
- *I. pseudobuxus* Reissek
- *I. taubertiana* Loes.
- *I. theezans* Reissek.

It has been reported that the taxonomic classification of *Ilex* species is workable by means of chemometrics and fingerprinting methods (Kim et al. 2010). This and other studies demonstrate the interest in the discrimination between different *Ilex* plants (Giberti 1979) with the aim of preserving quality and integrity (authenticity) of the real *yerba mate*. Some of the studied features concern the substantial absence of caffeine and different amounts of saponins in non-*yerba mate* leaves if compared with the authentic *I. paraguariensis* leaf.

Some doubts about the exact definition of *yerba mate* in the past occurred in past years because of the similarity between *I. paraguariensis* and *I. argentina* (found in the area approximately located between *Santa Cruz de la Sierra,* Bolivia, and *Andagalá*, Argentina). As a result, *yerba mate* was initially defined as a native Bolivian plant, until recent times. This confusion depends on the remarkable diversification of the *Ilex* genus (more than 600 species) and its localisation, not only in South America but also in East Asia (Croge et al. 2020).

References

Botanical-Online SL (2019) Mate plant. Botanical-online SL. Available: https://www.botanical-onl ine.com/en/botany/mate-plant. Accessed 14 Dec 2020

Choi YH, Sertic S, Kim HK, Wilson EG, Michopoulos F, Lefeber AW, Erkelens C, Prat Kricun SD, Verpoorte R (2005) Classification of Ilex species based on metabolomic fingerprinting using nuclear magnetic resonance and multivariate data analysis. J Agric Food Chem 53(4):1237–1245. https://doi.org/10.1021/jf0486141

Croge CP, Cuquel FL, Pintro PTM (2020) Yerba mate: cultivation systems, processing and chemical composition. A review. Sci Agricola 78(5):e20190259. https://doi.org/10.1590/1678-992X-2019-0259

Giberti GC (1979) Las especies argentinas del género Ilex L. (Aquifoliaceae). Darwiniana 22, 1–3:217–240

Heck CI, De Mejia EG (2007) Yerba Mate Tea (Ilex paraguariensis): a comprehensive review on chemistry, health implications, and technological considerations. J Food Sci 72(9):R138–R151. https://doi.org/10.1111/j.1750-3841.2007.00535.x

Kim HK, Khan S, Wilson EG, Kricun SDP, Meissner A, Goraler S, Deelder AM, Choi YH, Verpoorte R (2010) Metabolic classification of South American Ilex species by NMR-based metabolomics. Phytochem 71(7):773–784. https://doi.org/10.1016/j.phytochem.2010.02.001

Matta FV (2019) Chemical analysis of typical beverages and Açaí Berry from South America dissertation. University of Surrey, Guildford

Chapter 3
Chemical Profiles of *Yerba Mate* Infusions Across South American Countries

Abstract From the chemical viewpoint, the compositional profile of different *yerba mate* products influences the resulting infusion, both in the hot and in the cold versions (*mate, chimarrão*, and *tererê*). The chemical composition of these products is always referred to hot and cold infusions instead of the dried products. Naturally, several factors influence the aqueous extraction of bioactive principles, and the quantitative composition has a notable importance. In relation to *Ilex paraguariensis* and related infusions, the main bioactive classes found so far are: polyphenols; xanthenes; saponins; mineral elements; and vitamins. In detail, the chemical profiles of *yerba mate* infusions can be defined 'peculiar' depending on the qualitative and quantitative estimation of phenolics (chlorogenic acid, gallic acid, quercetin, rutin, etc.), xanthines (caffeine, theobromine, theophylline, and trigonelline), and saponins (matesaponins 1–5). The remaining chemical classes may be interesting also (especially minerals), taking into account that temperature and water volumes can remarkably influence the solubility of bioactive principles in infusions.

Keywords Caffeine · Chemical profile · Chlorogenic acid · *Ilex paraguariensis* · Manganese · (+)-Matesaponin 1 · *Yerba mate*

Abbreviations

CAS   Chemical Abstract Service
IUPAC   International Union of Pure and Applied Chemistry

3.1 Chemical Profiles of *Yerba Mate* Infusions

From the chemical viewpoint, the compositional profile of different *yerba mate* products influences the resulting infusion, both in the hot and in the cold versions (generally named *mate, chimarrão*, and *tererê*) (Matta 2019). Anyway, the chemical composition of these products is always referred to hot and cold infusions instead

© The Author(s), under exclusive license to Springer Nature Switzerland AG 2021
C. Iommi, *Chemistry and Safety of South American Yerba Mate Teas*,
Chemistry of Foods, https://doi.org/10.1007/978-3-030-69614-6_3

of the dried products. Naturally, several factors influence the aqueous extraction of bioactive principles.

The following list shows the approximate quantitative composition, without percentages, of the main bioactive classes found in *yerba mate* infusions (Barbosa et al. 2015; Bastos et al. 2007; Heck and De Mejia 2007; Matta 2019):

(1) Polyphenols
(2) Xanthines (purine alkaloids)
(3) Saponins (a group of natural glycosides)
(4) Minerals
(5) Vitamins.

More in detail, this composition can be described with concern to (1) molecular classes and (2) the most important representative compounds per class, when possible (Botanical-Online 2019; Schneider 2017):

(a) Polyphenols. Main representative compounds: chlorogenic acid, gallic acid, 4,5-dicaffeoylquinic acid, quercetin, and rutin
(b) Xanthines (purine alkaloids). Main representative compounds: caffeine, theobromine, theophylline, and trigonelline
(c) Saponins (a group of natural glycosides)
(d) Minerals. Main representative compounds: potassium, magnesium, manganese iron, aluminium, and zinc
(e) Vitamins. Representative compounds: choline, riboflavin, vitamin C, panthotenic acid, and pyridoxine.

The following sections discuss in detail the quail–quantitative profiles of *yerba mate* infusions in relation to four groups: polyphenols; xanthines; saponins; and mineral trace elements (vitamins are excluded in this ambit).

Once more, it has to be remembered that variable results concerning *yerba mate* infusions have to be taken into account: the temperature and water volumes can remarkably influence the solubility of minerals and other bioactive principles.

3.1.1 Polyphenols

In general, the quantity of extractable polyphenols from *yerba mate* leaves depends on the quality of leaves themselves (Chaps. 1 and 2), the dimension of particles (the direct relationship between superficial area and extractability of water should be taken into account), and the possible mixture of different *yerba mate* selections. In the last situation, the possible fraud concerning addition or replacement of authentic *I. paraguariensis* leaves with *Ilex* spp non-*yerba mate* raw materials is not considered (Chap. 2).

It has been reported that the quantity of polyphenols should be around 92 mg equivalents of chlorogenic acid per each gram of dried, except for the possibility of homogeneous selections of *yerba mate* (Bastos et al. 2007; Matta 2019). The problem with polyphenols is also dependent, from the analytical angle, on the choice of extraction solvent(s): the most powerful solvent appears acetone. On the other side, what about the real amount of water-soluble polyphenols (Heck and De Mejia 2007; Turkmen et al. 2006), because their importance is related to the real consumption? Anyway, chlorogenic acid appears to be the most abundant polyphenol in *I. paraguariensis* infusions. On the other side, it has been reported that catechins (very common molecules in green tea) are substantially absent (Heck and De Mejia 2007). This information has been repeatedly confirmed in the scientific literature, although one single work reports that gallocatechin and epicatechin may jointly reach 26% of the total organic extract in *yerba mate* (Schneider 2017). However, this information is not confirmed by other researchers. Consequently, the absence of analytically detectable catechins in *yerba mate* has to be confirmed, also as a simple discrimination method if compared with green and black teas (Chandra and De Mejia 2004; Hara 2001; Heck and De Mejia 2007).

In fact, the extraction of chlorogenic acid has been reported (in association with caffeine) with a chloroform–methyl alcohol–water (for a nuclear magnetic resonance study), while the aqueous extraction appears a good option (Choi et al. 2005; Heckman et al. 2010). The abundance of chlorogenic acid has been repeatedly confirmed not only in *yerba mate* but also in other *Ilex* species (Filip et al. 2001; Heckman et al. 2010). With exclusive relation to *yerba mate*, this compound is reported to be approximately 42% of the organic extract (Schneider 2017). A selection of data from different papers allows knowing the amount of chlorogenic acid may notably vary from 3.66 to 50 mg/100 ml of *chimarrão* and *tererê* extracts (Brazilian infusions) to more than 900 mg/100 ml for roasted *mate*, while the maximum amount can arrive to 1207 mg/100 ml with concern to extracts of green leaves (Butiuk et al. 2016; Croge et al. 2020; Lima et al. 2016; Mateos et al. 2018; Meinhart et al. 2010, 2018; Riachi et al. 2018; Silveira et al. 2017). Substantially, the amount of this compound decreases from green leaves to the final product, but a number of variables—different selections, native or cultivated plants, harvesting methods, analytical choices, etc.—are able to influence the result. The only reliable information is that this polyphenol is the most abundant molecule for this class in *yerba mate*.

3.1.1.1 The Most Abundant Polyphenol in *Yerba Mate*: Chlorogenic Acid

Chlorogenic acid is an important polyphenolic compound. This molecule—other names: 3-*O*-caffeoylquinic acid and 3-(3,4-dihydroxycinnamoyl)quinic acid—has a molecular weight of 354.31 Da with molecular formula: $C_{16}H_{18}O_9$, and Chemical Abstract Service (CAS) Number: 327-97-9, is substantially an ester obtained by condensation between *trans*-caffeic acid and quinic acid. As a result, it has to be considered as a cinnamate ester and also as tannin (Pubchem 2020a). It is a member of the family of non-flavonoid phenolics, with caffeic and quinic acid esters, and related isomers including 3-*O*-caffeoylquinic acid, 3,5-dicaffeoylquinic acid, and 4,5-dicaffeoylquinic acid (Croge et al. 2020; Heck and De Mejia 2007). Peculiar

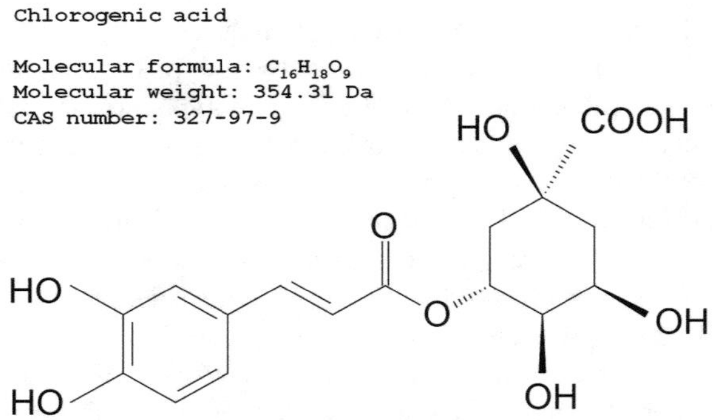

Chlorogenic acid

Molecular formula: $C_{16}H_{18}O_9$
Molecular weight: 354.31 Da
CAS number: 327-97-9

Fig. 3.1 Chlorogenic acid is an ester obtained by condensation between *trans*-caffeic acid and quinic acid. As a result, it has to be considered as a cinnamate ester and also as a tannin (Pubchem 2020a). BKchem version 0.13.0, 2009 (https://bkchem.zirael.org/index.html), has been also used for drawing this structure

chemical properties are discussed in Chap. 4 when speaking of antioxidant power and other safety features. The structure of chlorogenic acid is displayed in Fig. 3.1.

3.1.2 Xanthines

Xanthines are normally found in tea, chocolate products, coffee, and *yerba mate* also (Heck and De Mejia 2007). With reference to *I. paraguariensis*, the most abundant xanthines are reported to be caffeine and theobromine, although other xanthenes may be detected (caffeic acid, caffeoylshikimic acid, several caffeoyl derivatives, kaempferol, quercetin, quinic acid, and rutin). The matter of caffeine is interesting if *yerba mate* is compared with coffee: 260 or more mg of caffeine per day may be obtained by *mate* infusions consumed in South America because the normal *mate* consumption is based on approximately 500 ml of aqueous infusion. Moreover, the real caffeine amount in *yerba mate* raw materials has to be higher ($\geq$30%) than reported values because processing methods (Chap. 1) can cause the notable decrease of this xanthine, especially in the *sapeco* and *barbaqua* steps. Anyway, the repeated infusion and consumption cycle can explain the reason which caffeine is still a high value. The relevant loss of moisture can notably increase dry matters, and consequently water extraction and daily consumption (Heck and De Mejia 2007). A selection of data from different papers allows knowing the amount of caffeine may notably vary from 6.30 to 68.30 mg/100 ml of *chimarrão* and *tererê* extracts (Brazilian infusions), while the maximum amount in extracts from green leaves may reach 47% of the maximum quantity in Brazilian infusions (Croge et al. 2020; Friedrich et al. 2017; Gebara et al. 2017; Heck and De Mejia 2007; Holowaty et al. 2016; Konieczynski et al. 2017). Theobromine, another methylxanthine, is reported to slightly exceed 4 mg/100 ml in Brazilian infusions. Consequently, this compound

Caffeine (1,3,7-trimethylxanthine)

Molecular formula:$C_8H_{10}N_4O_2$
Molecular weight: 194.19 Da
CAS number: 58-08-2

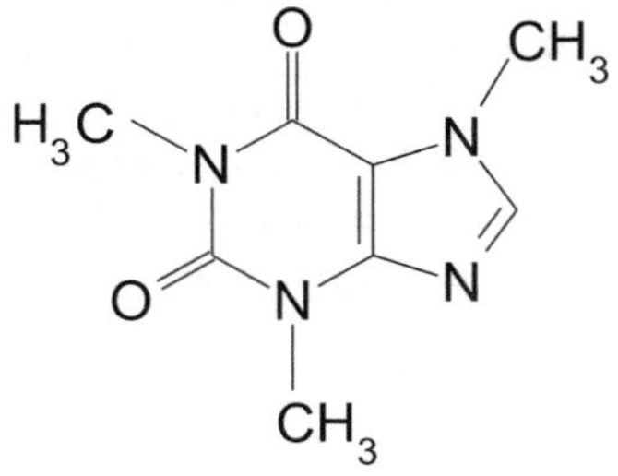

Fig. 3.2 Caffeine, also named mateine or guaranine (found in *yerba mate* and *guaranà*, respectively), is an important xanthine. This compound is well known because of its activity as a central nervous system stimulant in humans. BKchem version 0.13.0, 2009 (https://bkchem.zirael.org/index.html), has been also used for drawing this structure

(and also theophylline) may be relevant enough without the same intake of caffeine. Interestingly, caffeine is also intended to be 'mateine' when speaking of *yerba mate* (Burdan 2015; Martínez-Huitle et al. 2010; Riahi et al. 2009; Sanchis-Gomar et al. 2015; Senchina et al. 2014).

3.1.2.1 The Most Abundant Xanthine in *Yerba Mate*: Caffeine

Caffeine, also named mateine or guaranine (if found in *yerba mate* and *guaranà*, respectively), is an important xanthine. This molecule—official IUPAC[1] name: 1,3,7-trimethylxanthine; other names: 5-*O*-caffeoyl quinic acid and 3,7-dihydro-1,3,7-trimethyl-1*H*-purine-2,6-dione—has a molecular weight of 194.19 Dalton, with molecular formula $C_8H_{10}N_4O_2$ and Chemical Abstract Service (CAS) Number: 58-08-2. Caffeine is well known because of its activity as a central nervous system stimulant in humans. It is found in notable amounts when speaking of coffee, tea, chocolate, and also *yerba mate*. Its biological function in plants is related to natural pesticide action (Pubchem 2020b). Other peculiar chemical properties are discussed in Chap. 4. The chemical structure of caffeine is displayed in Fig. 3.2.

3.1.3 Saponins

The importance of saponins in *yerba mate* and other plants such as ginseng (*Panax ginseng*) roots is linked to their surfactant ability to disrupt vegetable membranes,

---

[1] International Union of Pure and Applied Chemistry (IUPAC).

being so able to create large micelles with bile acids and steroids. This fact may explain, in part at least, certain features of folk medicines.

In addition, their notable bitterness and the high solubility in water seem to explain partially the distinct organoleptic features of *yerba mate* infusions. Chemically, these compounds are subdivided into steroidal and triterpenoid saponins, depending on the aglycone skeleton (Heck and De Mejia 2007). In relation to *yerba mate*, the most interesting (and perhaps promising) saponins are reported to be matesaponins 1, 2, 3, 4, and 5. Interestingly, the water solubility of these matesaponins is reported to be close to 100% when speaking of matesaponin 1, while the total amount of extractable saponins may slightly exceed 23 μg/mL for one gram of dried leaves in 100 mL water (Heck and De Mejia 2007).

3.1.3.1 The Most Abundant Saponin in *Yerba Mate*: (+)-Matesaponin 1

(+)-Matesaponin 1 (Fig. 3.3) is the most abundant saponin in *yerba mate*. It is also

(+)-Matesaponin 1

Molecular formula: $C_{47}H_{76}O_{17}$
Molecular weight: 913.1 Da
CAS number: 126622-38-6

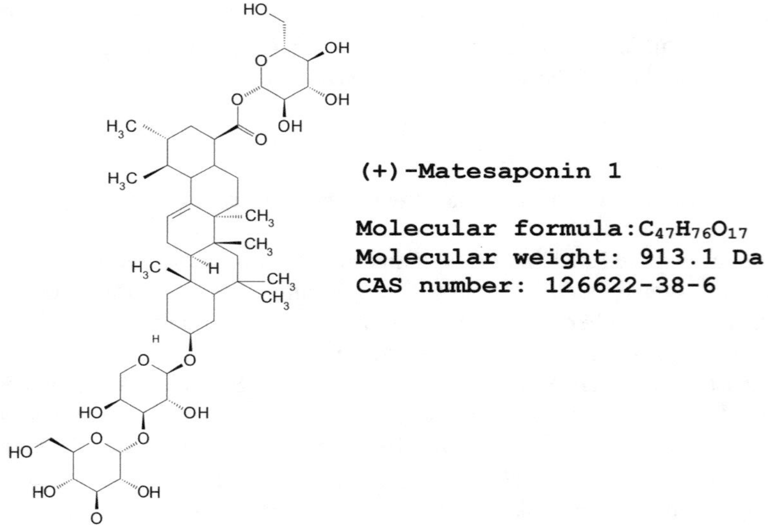

Fig. 3.3 (+)-Matesaponin 1 is the most abundant saponin in *yerba mate*. It is also found in other *Ilex* species. This molecule has a molecular weight of 913.1 Da, with molecular formula $C_{47}H_{76}O_{17}$ (Pubchem 2020c), and it is reported (with other matesaponins) to show hypocholesteremic effects and wish some anti-cancer and anti-parasitic property (Taketa et al. 2004a; b). BKchem version 0.13.0, 2009 (https://bkchem.zirael.org/index.html), has been also used for drawing this structure

found in other *Ilex* species. This molecule[2] has a molecular weight of 913.1 Da, with molecular formula $C_{47}H_{76}O_{17}$ (Pubchem 2020c). (+)-Matesaponin 1 and other triterpenoid saponins are reported to show hypocholesteremic effects and wish some anti-cancer and anti-parasitic property (Taketa et al. 2004a, b).

3.1.4 Trace Mineral Substances

The mineral composition of extractable *yerba mate* strongly depends on the following factors (Barbosa et al. 2015; Bastos et al. 2007; Croge et al. 2020):

(a) Agricultural practices and seasons, including used fertilisers, if any
(b) The difference between mineral extraction by *yerba mate* roots and the correspondent storage of extracted mineral in leaves. As an example, calcium and sodium show opposite behaviours, probably because of different aqueous solubility.

In general, and with exclusive reference to *yerba mate* leaves, the following mineral elements are present in remarkable amounts, also if compared with other vegetable products used for infusion purposes, and with the exclusion of non-trace elements calcium and magnesium (Bragança et al. 2011; Matta 2019; Rusinek-Prystupa et al. 2016):

(1) Manganese (over 260 mg/kg, with a maximum amount exceeding 1000 or 2200 mg/kg, dry weight, depending on references; or >2 mg/l of infusion)
(2) Zinc (>64 mg/kg, dry weight, or >0.4 mg/l of infusion)
(3) Iron (>32 mg/kg, dry weight, or >0.1 mg/l of infusion)
(4) Copper (>5 mg/kg, dry weight, or >0.03 mg/l of infusion).

The revealed differences, with the exclusion of expected calcium and magnesium quantities, concern mainly manganese. Interestingly, calcium and magnesium have remarkable amounts in spite of the nature of high-acidity soils. It has to be considered that manganese absorption is probably influenced by the following factors (Matta 2019):

(a) *Yerba mate* leaves contain more manganese and other minor elements if collected in organic or traditional organic plantations (instead of native plants).
(b) The peculiar product features (green VS roasted; loose VS teabag packages) seem to be critical. In detail, roasted and/or teabag-packaged *yerba mate* products appear to have more manganese and other minor elements. In contrast, the country of origin (e.g. Argentina VS Brazil) does not seem to influence the mineral profile.

---

[2]Official IUPAC name: [(2S,3R,4S,5S,6R)-3,4,5-trihydroxy-6-(hydroxymethyl)oxan-2-yl] (1S,2R,4aS,6aR,6aS,6bR,10S,12aR,14bR)-10-[(2S,3R,4S,5S)-3,5-dihydroxy-4-[(2R,3R,4S,5S,6R)-3,4,5-trihydroxy-6-(hydroxymethyl)oxan-2-yl]oxyoxan-2-yl]oxy-1,2,6a,6b,9,9,12a-heptamethyl-2,3,4,5,6,6a,7,8,8a,10,11,12,13,14b-tetradecahydro-1H-picene-4a-carboxylate. IUPAC is for: International Union of Pure and Applied Chemistry.

The above-mentioned data have to be considered when speaking of mineral intake values per day. In fact, each trace mineral element (excluding calcium and magnesium) seems to be absorbed by humans with a maximum intake which should be less than 5%. The only exception between trace elements is manganese, probably because of its really remarkable quantity if compared with zinc, iron, and copper.

Interestingly, the general quantity of extractable minerals in a *mate* infusion seems to augment if tannins decrease.

References

Barbosa JZ, Zambon LM, Motta ACV, Wendling I (2015) Composition, hot-water solubility of elements and nutritional value of fruits and leaves of yerba mate. Ciência E Agrotecnologia 39(6):593–603. https://doi.org/10.1590/S1413-70542015000600006

Bastos DHM, De Oliveira DM, Matsumoto RT, Carvalho PDO, Ribeiro ML (2007) Yerba mate: pharmacological properties, research and biotechnology. Med Aromat Plant Sci Biotechnol 1(1):37–46

Botanical-Online SL (2019) Mate plant. Botanical-online SL. Available https://www.botanical-online.com/en/botany/mate-plant. Accessed 14 Dec 2020

Bragança VLC, Melnikov P, Zanoni LZ (2011) Trace elements in different brands of Yerba Mate Tea. Biol Trace Elem Res 144:1197–1204. https://doi.org/10.1007/s12011-011-9056-3

Burdan F (2015) Caffeine in Coffee. In: Preedy VR (ed) Coffee in health and disease prevention. Academic Press, Cambridge, pp 201–207. https://doi.org/10.1016/B978-0-12-409517-5.00022-X

Butiuk AP, Martos MA, Adachi O, Hours RA (2016) Study of the chlorogenic acid content in yerba mate (Ilex paraguariensis St. Hil.): effect of plant fraction, processing step and harvesting season. J Appl Res Med Aromat Plants 3(1):27–33. https://doi.org/10.1016/j.jarmap.2015.12.003

Chandra S, De Mejia GE (2004) Polyphenolic compounds, antioxidant capacity, and quinone reductase activity of an aqueous extract of Ardisia compressa in comparison to Mate (Ilex paraguariensis) and green (Camellia sinensis) teas. J Agric Food Chem 52(11):3583–3590. https://doi.org/10.1021/jf0352632

Choi YH, Sertic S, Kim HK, Wilson EG, Michopoulos F, Lefeber AW, Erkelens C, Prat Kricun SD, Verpoorte R (2005) Classification of Ilex species based on metabolomic fingerprinting using nuclear magnetic resonance and multivariate data analysis. J Agric Food Chem 53(4):1237–1245. https://doi.org/10.1021/jf0486141

Croge CP, Cuquel FL, Pintro PTM (2020) Yerba mate: cultivation systems, processing and chemical composition. A review. Sci Agricola 78(5):e20190259. https://doi.org/10.1590/1678-992X-2019-0259

Filip R, Lopez P, Giberti G, Coussio J, Ferraro G (2001) Phenolic compounds in seven South American Ilex species. Fitoter 72(7):774–778. https://doi.org/10.1016/S0367-326X(01)00331-8

Friedrich JC, Gonela A, Vidigal MCG, Vidigal-Filho PS, Sturion JA, Cardozo-Junior EL (2017) Genetic and phytochemical analysis to evaluate the diversity and relationships of mate (Ilex paraguariensis A. ST.-HIL.): elite genetic resources in a germplasm collection. Chem Biodivers14:e1600177. https://doi.org/10.1002/cbdv.201600177

Gebara KS, Gasparotto-Junior A, Santiago PG, Cardoso CAL, Souza LM, Morand C, Costa TA, Cardozo-Junior EL (2017) Daily intake of chlorogenic acids from consumption of maté (Ilex paraguariensis A. St.- Hil.) traditional beverages. J Agric Food Chem 65(46):10093–10100. https://doi.org/10.1021/acs.jafc.7b04093

Hara Y (2001) Green tea: health benefits and applications. Marcel Dekker Inc., New York, NY, pp 16–20

Heck CI, De Mejia EG (2007) Yerba Mate Tea (Ilex paraguariensis): a comprehensive review on chemistry, health implications, and technological considerations. J Food Sci 72(9):R138–R151. https://doi.org/10.1111/j.1750-3841.2007.00535.x

Heckman MA, Weil J, De Mejia EG (2010) Caffeine (1, 3, 7-trimethylxanthine) in foods: a comprehensive review on consumption, functionality, safety, and regulatory matters. J Food Sci 75(3):R77–R87. https://doi.org/10.1111/j.1750-3841.2010.01561.x

Holowaty SA, Trela V; Thea AE, Scipioni GP, Schmalko ME (2016) Yerba maté (Ilex paraguariensis st. Hil.): chemical and physical changes under different aging conditions. J Food Proc Eng 39(1):19–30. https://doi.org/10.1111/jfpe.12195

Konieczynski P, Viapiana A, Wesolowski M (2017) Comparison of infusions from black and green teas (Camellia sinensis L. Kuntze) and erva-mate (Ilex paraguariensis A. St.-Hil.) based on the content of essential elements, secondary metabolites, and antioxidant activity. Food Anal Methods 10:3063–3070. https://doi.org/10.1007/s12161-017-0872-8

Lima JP, Farah A, King B, Paulis T, Martin PR (2016) Distribution of major chlorogenic acids and related compounds in Brazilian green and toasted Ilex paraguariensis (maté) leaves. J Agric Food Chem 64(11):2361–2370. https://doi.org/10.1021/acs.jafc.6b00276

Martínez-Huitle CA, Fernandes NS, Ferro S, De Battisti A, Quiroz MA (2010) Fabrication and application of Nafion®-modified boron-doped diamond electrode as sensor for detecting caffeine. Diam Relat Mater 19(10):1188–1193. https://doi.org/10.1016/j.diamond.2010.05.004

Mateos R, Baeza G, Sarriá B, Bravo L (2018) Improved LC-MSn characterisation of hydroxycinnamic acid derivatives and flavonols in different commercial mate (Ilex paraguariensis) brands. Quantification of polyphenols, methylxanthines, and antioxidant activity. Food Chem 241:232–241. https://doi.org/10.1016/j.foodchem.2017.08.085

Matta FV (2019) Chemical analysis of typical beverages and Açaí Berry from South America dissertation. University of Surrey, Guildford

Meinhart AD, Bizzotto CS, Ballus CA, Rybka ACP, Sobrinho MR, Cerro-Quintana RS, Teixeira-Filho J, Godoy HT (2010) Methylxanthines and phenolics content extracted during the consumption of mate (Ilex paraguariensis St. Hil) beverages. J Agric Food Chem 58(4):2188–2193. https://doi.org/10.1021/jf903781w

Meinhart AD, Caldeirão L, Damin FM, Filho JT, Godoy HT (2018) Analysis of chlorogenic acids isomers and caffeic acid in 89 herbal infusions (tea). J Food Comp Anal 73:76–82. https://doi.org/10.1016/j.jfca.2018.08.001

Pubchem (2020a) Chlorogenic acid. National Center for Biotechnology Information, Bethesda, MD. Available https://pubchem.ncbi.nlm.nih.gov/compound/Chlorogenic-acid. Accessed 14 Dec 2020

Pubchem (2020b) Caffeine. National Center for Biotechnology Information, Bethesda, MD. Available https://pubchem.ncbi.nlm.nih.gov/compound/2519. Accessed 14 Dec 2020

Pubchem (2020c) (+)-Matesaponin 1. National Center for Biotechnology Information, Bethesda, MD. Available https://ncbi.nlm.nih.gov/compound/Matesaponin-1. Accessed 14 Dec 2020

Riachi LG, Simas DLR, Coelho GC, Marcellini PS, Silva AJRS, de Maria CAB (2018) Effect of light intensity and processing conditions on bioactive compounds in maté extracted from yerba mate (Ilex paraguariensis A. St.-Hil.). Food Chem 266:317–322. https://doi.org/10.1016/j.foodchem.2018.06.028

Riahi S, Faridbod F, Ganjali MR (2009) Caffeine sensitive electrode and its analytical applications. Sens Lett 7(1):42–49. https://doi.org/10.1166/sl.2009.1008

Rusinek-Prystupa E, Marzec Z, Sembratowicz I, Samolińska W, Kiczorowska B, Kwiecień M (2016) Content of selected minerals and active ingredients in teas containing yerba mate and rooibos. Biol Trace Elem Res 172:266–275. https://doi.org/10.1007/s12011-015-0588-9

Sanchis-Gomar F, Pareja-Galeano H, Cervellin G, Lippi G, Earnest CP (2015) Energy drink overconsumption in adolescents: implications for arrhythmias and other cardiovascular events. Can J Cardiol 31(5):572–575. https://doi.org/10.1016/j.cjca.2014.12.019

Schneider M (2017) Determinação da adulteração da erva-mate por adição de sacarose empregando espectroscopia no infravermelho (atr-ftir) em conjunto com ferramentas quimiométricas. Dissertation, Universidade Federal do Rio Grande do Sul, Porto Alegre

Senchina DS, Hallam JE, Kohut ML, Nguyen NA, Perera MADN (2014) Alkaloids and athlete immune function: caffeine, theophylline, gingerol, ephedrine, and their congeners. Exerc Immunol Rev 20:68–93

Silveira TFF, Meinhart AD, Souza TCL, Cunha ECE, Moraes MR, Godoy HT (2017) Chlorogenic acids and flavonoid extraction during the preparation of yerba mate based beverages. Food Res Int 102:348–354. https://doi.org/10.1016/j.foodres.2017.09.098

Taketa ATC, Breitmaier E, Schenkel EP (2004) Triterpenes and triterpenoidal glycosides from the fruits of Ilex paraguariensis (Mate). J Braz Chem Soc 15:205–211. https://doi.org/10.1590/S0103-50532004000200008

Taketa ATC, Gnoatto SCB, Gosmann G, Pires VS, Schenkel EP, Guillaume D (2004) Triterpenoids from Brazilian Ilex species and their in vitro antitrypanosomal activity. J Nat Prod 67:1697–1700. https://doi.org/10.1021/np040059+

Turkmen N, Sari F, Velioglu YS (2006) Effects of extraction solvents on concentration and antioxidant activity of black and black Mate tea polyphenols determined by ferrous tartrate and Folin-Ciocalteu methods. Food Chem 99:835–841. https://doi.org/10.1016/j.foodchem.2005.08.034

Chapter 4
Bioactive Features of *Yerba Mate* Infusions: Alkaloids, Phenolics, and Other Stimulating Compounds

Abstract This chapter concerns the correct interpretation of many claimed safety properties ascribed to *Ilex paraguariensis* (*yerba mate*) from the safety viewpoint, on the basis of the chemical composition of different *yerba mate* products. The daily intake of active principles is important enough, and the qualitative identification of these molecules is critical, taking into account that *yerba mate* is consumed as hot and the cold infusions (generally named *mate*, *chimarrão*, and *tererê*). The chemical profile of these infusions comprehends polyphenols, xanthines (purine alkaloids), saponins (natural glycosides), minerals and trace elements, and also vitamins. The main health properties ascribed to *yerba mate* may be correlated to these categories and main representative active principles, with the exclusion of vitamins which can be found in a plethora of non-*I. paraguariensis* vegetables.

Keywords Anti-inflammatory · Antioxidant · Caffeine · Chlorogenic acid · *Ilex paraguariensis* · Manganese · *Yerba mate*

Abbreviations

Boron	B
Cobalt	Co
Copper	Cu
Iodine	I
Iron	Fe
Manganese	Mn
Molybdenum	Mo
Reactive oxygen species	ROS
Zinc	Zn

4.1   Safety and Health Features of *Yerba Mate*

From the safety viewpoint, the chemical composition of different *yerba mate* products can be considered as the key for the correct interpretation of many claimed safety properties ascribed to *I. paraguariensis*. In this ambit, the daily intake of active principles is important enough, but the qualitative identification of these molecules is critical (also ins relation to the identification of chemical classes and sub-typologies). In addition, it has to be remembered that *yerba mate* is not consumed 'as it is': all available researches carried out so far are always related to hot and cold infusions (generally named *mate*, *chimarrão*, and *tererê*) (Matta 2019).

As discussed in Chap. 3, the approximate quantitative profile of the main bioactive classes found in *yerba mate* infusions (Barbosa et al. 2015; Bastos et al. 2007; Botanical-Online SL 2019; Heck and De Mejia 2007; Matta 2019; Schneider 2017) should comprehend the following groups:

(a)   Polyphenols. In relation to this broad group, the main representative compounds in *yerba mate* are: chlorogenic acid, gallic acid, 4,5-dicaffeoylquinic acid, quercetin, and rutin.
(b)   Xanthines (purine alkaloids). In this group, the most important active principles are (in descending order): caffeine, theobromine, theophylline, and trigonelline.
(c)   Saponins (natural glycosides). In this class, the most studied molecules are five matesaponins.
(d)   Minerals. In detail, and with the exclusion of calcium and magnesium, the main representative of trace elements seems to be manganese.
(e)   Vitamins. In detail, the following vitamins have been found in *yerba mate*: panthotenic acid, choline, riboflavin, pyridoxine, and vitamin C.

The following sections discuss the main health properties correlated with these categories and main representative active principles, with the exclusion of vitamins which can be found in a plethora of non-*I. paraguariensis* vegetable species.

4.2   Health and Polyphenols

In general, polyphenols and minerals are claimed to have important functions when speaking of human health (and, naturally, human nutrition as a synergic and collateral therapy for human diseases). In relation to *yerba mate*, the popularity and the traditional consumption of hot and cold infusions have been always accompanied with folk histories and claimed health effects (Matta 2019). In general, these properties are correlated with antioxidant power, and polyphenols are certainly one of the most promising natural substances with similar features (Barbera 2020; Barbieri et al. 2019; Bhagat et al. 2019; Delgado et al. 2016, 2019; Haddad and Parisi 2020; Haddad et al. 2020a, b; Issaoui et al. 2020; Matta 2019; Parisi 2016, 2019, 2020;

Parisi and Dongo 2020; Parisi and Haddad 2019; Parisi et al. 2020; Singla et al. 2019).

Phenolics are chemically a broad group of molecules with many sub-classes: the presence of one or more aromatic rings and the concomitant presence of one or more hydroxyl groups on the same rings as conjugated structures determine the role of partner in oxidation and reduction reactions. As a result, vegetable organisms can use polyphenols for different reasons, including (Matta 2019):

(a)   Anti-insect action
(b)   Anti-mammal feeding substance
(c)   Colour enhancement
(d)   Defence against ultraviolet rays
(e)   Defence against thermal variations
(f)   Antioxidant activity (with the aim of preserving extremely important molecules and tissues)
(g)   Defence against toxic heavy metals by means of chelation effect
(h)   General defence against pathogen agents.

In this ambit, *yerba mate* is reported to block chemically reactive oxygen species (ROS) by means of the enhancement of peroxidase-like activity. This biological 'power' is notably augmented on condition that polyphenols increase, meaning substantially that vegetable polyphenols can become important 'partners' of the natural antioxidant systems. In other terms, the augment of polyphenols 'boosts' the performance of enzymatic systems and reactions relying on oxidation–reduction reactions. As a consequence, polyphenols appear to justify the claimed *mate* action against different diseases and safety-menacing facts, including (a) oxidative stress in livers and heart, (b) general cardiovascular diseases and post-ischemic reperfusion and tissutal damages after heart attack, (c) peroxidation of lipids, (d) damages to the genetic human code, (e) stroke and myocardial ischemia, and (f) cellular death. In some case, *mate* consumption appears to be more significant than the assumption of red wine and green tea (Gorgen et al. 2005; Heck and De Mejia 2007; Schinella et al. 2005).

In this ambit, chlorogenic acid seems to be the main agent, in *yerba mate* and in other vegetables (Heck and De Mejia 2007; Ristow 2014). Actually, there are different chlorogenic acids, including 5-caffeoylquinic acid (Matta 2019). In general, chlorogenic acid is reported to have different health effects against pathogens, diabetes, cancer, and anti-inflammatory properties (against oxidative stress) also because of its action as a metal-chelating agent, and the reduction of lipid peroxidation. On the other time, more research is needed when speaking of possible countereffects such as the inhibition of endogenous antioxidant systems (Tajik et al. 2017). Anyway, the abundance of chlorogenic acid and its similar compounds has to be highlighted as one of the most interesting health features ascribed to *yerba mate* infusions.

4.3   Health and Xanthines

As discussed in Chap. 3, xanthines are normally found in tea, chocolate products, coffee, and *yerba mate* (Heck and De Mejia 2007). Their health effects are generally correlated and well represented by the most interesting of these molecules, caffeine. Its effects can be simply listed as follows:

(a)   Antioxidant agent (with polyphenols and some minerals, caffeine and other xanthines may be an excellent barrier against many oxidant menaces)
(b)   Vasodilatator agent (this reason explains very well its consumption as coffee and coffee products. In relation to *yerba mate*, the reason is still valid)
(c)   Anti-carcinogenic agent
(d)   Stimulant molecule (its use in many non-alcoholic beverages is linked to its stimulant capability)
(e)   Diuretic agent
(f)   Anti-obesity factor (weight loss).

It has to be remembered that caffeine is also intended to be 'mateine' when speaking of *yerba mate*. With reference to another South American beverage, *guaranà*, the same molecule is also named 'guaranine' demonstrating that the strong interest in different caffeine properties is not only a distinct feature of coffee products (Burdan 2015; Martínez-Huitle et al. 2010; Riahi et al. 2009; Sanchis-Gomar et al. 2015; Senchina et al. 2014).

4.4   Health and Saponins

The main chemical feature of saponins is the high solubility in water. This fact can explain the related surfactant ability to disrupt vegetable membranes, being so able to create large micelles with bile acids and steroids. On the other hand, certain features of folk medicines associated with (or containing) *yerba mate* may partially be explained, with specific reference to claimed hypocholesterolemic feature of *I. paraguariensis*. In other words, substances such as the five matesaponins (1–5) found in *yerba mate* may be able to inhibit the passive diffusion of cholic acid. As a result, the production of dangerous micelles (which could not be absorbed and consequently eliminated by the human organism) would be avoided or limited. Saponins are also reported to have certain anti-fungal, antibacterial, and anti-parasitic activity among all reported features for *yerba mate* and other vegetable species (Ferreira et al. 1997; Fleck et al. 2019; Wang et al. 2010).

4.5 Health and Trace Mineral Substances: The Role of Manganese

In general, minerals and polyphenols are claimed to have important functions when speaking of human health and, naturally, human nutrition as a synergic and collateral therapy for human diseases (Nielsen and Hunt 1989). In relation to *yerba mate*, the popularity and the traditional consumption of hot and cold infusions have been always accompanied with folk histories and claimed health effects (Matta 2019). In general, these properties are correlated with antioxidant power (polyphenols) and the presence of trace micronutrients. The list of 'essential' trace elements with concentration between 0.1 and 100 mg/l—boron (B), cobalt (Co), copper (Cu), iodine (I), iron (Fe), manganese (Mn), molybdenum (Mo), and zinc (Zn)—has an important function when speaking of enzymatic reactions (Matta 2019). The lack of one or more than above-mentioned elements can surely cause deficiency diseases (Matta 2019).

In this ambit (trace elements only), manganese appears the most abundant and interesting metal in *yerba mate* products. Its biological functions into several enzymes include regulation of sugars in the blood flow, blood clotting, digestive processes, growth of bones, respiration (electron transport in plant leaves), and immunity responses (Matta 2019). Its presence is chemically linked to its oxidation states (minimum: $+2$, also the most stable and available in nature; maximum: $+7$, only in the permanganate ion). An interesting oxidation state ($+4$) is found in complexes, and several compounds with biological activity are natural complexes. The same presence of Mn in aqueous suspensions is generally influenced by conditions because manganese is easily oxidizable and reducible, with important consequences on biological processes in complexes. Anyway, Mn is water-soluble only if pH is acidic (<6). At present, a recommendation concerning Mn dietary intake levels for men and women is available with the following advices: 2.3 and 1.8 mg per day, respectively. Because of possible toxic effects (manganism) at elevated concentrations, the tolerable upper intake amount is 11 mg per day (Matta 2019).

References

Barbera M (2020) Reuse of food waste and wastewater as a source of polyphenolic compounds to use as food additives. J AOAC Int 103(4):906–914. https://doi.org/10.1093/jaocint/qsz025

Barbieri G, Bergamaschi M, Saccani G, Caruso G, Santangelo A, Tulumello T, Vibhute B, Barbieri G (2019) Processed meat and polyphenols: opportunities, advantages, and difficulties. J AOAC Int 102(5):1401–1406. https://doi.org/10.1093/jaoac/102.5.1401

Barbosa JZ, Zambon LM, Motta ACV, Wendling I (2015) Composition, hot-water solubility of elements and nutritional value of fruits and leaves of yerba mate. Ciênc Agrotec 39(6):593–603. https://doi.org/10.1590/S1413-70542015000600006

Bastos DHM, De Oliveira DM, Matsumoto RT, Carvalho PDO, Ribeiro ML (2007) Yerba mate: pharmacological properties, research and biotechnology. Med Aromat Plant Sci Biotechnol 1(1):37–46

Bhagat AR, Delgado AM, Issaoui M, Chammem N, Fiorino M, Pellerito A, Natalello S (2019) Review of the role of fluid dairy in delivery of polyphenolic compounds in the diet: chocolate milk, coffee beverages, Matcha green tea, and beyond. J AOAC Int 102(5):1365–1372. https://doi.org/10.1093/jaoac/102.5.1365

Botanical-Online SL (2019) Mate plant. Botanical-Online SL. Available https://www.botanical-online.com/en/botany/mate-plant. Accessed 14 Dec 2020

Burdan F (2015) Caffeine in coffee. In: Preedy VR (ed) Coffee in health and disease prevention. Academic Press, Cambridge, pp 201–207. https://doi.org/10.1016/B978-0-12-409517-5.00022-X

Delgado AM, Vaz de Almeida MD, Barone C, Parisi S (2016) Leguminosas na Dieta Mediterrânica—Nutrição, Segurança, Sustentabilidade. In: CISA—VIII Conferência de Inovação e Segurança Alimentar, ESTM-IPLeiria, Peniche

Delgado AM, Issaoui M, Chammem N (2019) Analysis of main and healthy phenolic compounds in foods. J AOAC Int 102(5):1356–1364. https://doi.org/10.1093/jaoac/102.5.1356

Ferreira F, Vázquez A, Güntner C, Moyna P (1997) Inhibition of the passive diffusion of cholic acid by the *Ilex paraguariensis* St. Hil saponins. Phytother Res 11:79–81. https://doi.org/10.1002/(SICI)1099-1573(199702)11:1%3C79::AID-PTR34%3E3.0.CO;2-R

Fleck JD, Betti AH, Da Silva FP, Troian EA, Olivaro C, Ferreira F, Verza SG (2019) Saponins from *Quillaja saponaria* and *Quillaja brasiliensis*: particular chemical characteristics and biological activities. Molecules 24(1):171. https://doi.org/10.3390/molecules24010171

Gorgen M, Turatti K, Medeiros AR, Buffon A, Bonan CD, Sarkis JJ, Pereira GS (2005) Aqueous extract of *Ilex paraguariensis* decreases nucleotide hydrolysis in rat blood serum. J Ethnopharmacol 97:73–77. https://doi.org/10.1016/j.jep.2004.10.015

Haddad MA, Parisi S (2020) The next big HITS. New Food Mag 23(2):4

Haddad MA, Dmour H, Al-Khazaleh JFM, Obeidat M, Al-Abbadi A, Al-Shadaideh AN, Al-Mazra'awi MS, Shatnawi MA, Iommi C (2020a) Herbs and medicinal plants in Jordan. J AOAC Int 103(4):925–929. https://doi.org/10.1093/jaoac/qsz026

Haddad MA, El-Qudah J, Abu-Romman S, Obeidat M, Iommi C, Jaradat DSM (2020b) Phenolics in Mediterranean and Middle East important fruits. J AOAC Int 103(4):930–934. https://doi.org/10.1093/jaocint/qsz027

Heck CI, De Mejia EG (2007) Yerba Mate Tea (*Ilex paraguariensis*): a comprehensive review on chemistry, health implications, and technological considerations. J Food Sci 72(9):R138–R151. https://doi.org/10.1111/j.1750-3841.2007.00535.x

Issaoui M, Delgado AM, Caruso G, Micali M, Barbera M, Atrous H, Ouslati A, Chammem N (2020) Phenols, flavors, and the Mediterranean diet. J AOAC Int 103(4):915–992. https://doi.org/10.1093/jaocint/qsz018

Martínez-Huitle CA, Fernandes NS, Ferro S, De Battisti A, Quiroz MA (2010) Fabrication and application of Nafion®-modified boron-doped diamond electrode as sensor for detecting caffeine. Diam Relat Mater 19(10):1188–1193. https://doi.org/10.1016/j.diamond.2010.05.004

Matta FV (2019) Chemical analysis of typical beverages and Açaí Berry from South America. Dissertation, University of Surrey, Guildford

Nielsen FH, Hunt JR (1989) Trace elements emerging as important in human nutrition. In: Proceedings of the fourteenth national databank conference, June 1989. Univ. Iowa, Iowa City, pp 135–143

Parisi S (2016) The world of foods and beverages today: globalization, crisis management and future perspectives. Learning.ly/The Economist Group. Available https://learning.ly/products/the-world-of-foods-and-beverages-today-globalization-crisis-management-and-future-perspectives. Accessed 09 Dec 2020

Parisi S (2019) Analysis of major phenolic compounds in foods and their health effects. AOAC J 102(5):1354–1355. https://doi.org/10.5740/jaoacint.19-0127

Parisi S (2020) Characterization of major phenolic compounds in selected foods by the technological and health promotion viewpoints. J AOAC Int 103(4):904–905. https://doi.org/10.1093/jaoacint/qsaa011

Parisi S, Dongo D (2020). Polifenoli e salute. I vegetali amici del sistema immunitario. Great Italian Food Trade. Available https://www.greatitalianfoodtrade.it/salute/polifenoli-e-salute-i-vegetali-amici-del-sistema-immunitario. Accessed 09 Dec 2020

Parisi S, Haddad MA (2019) Food safety 101. Al-Balqa Applied University, Al-Salt

Parisi S, Dongo D, Parisi C (2020) Resveratrolo, conoscenze attuali e prospettive. Great Italian Food Trade. Available www.greatitalianfoodtrade.it/integratori/resveratrolo-conoscenze-attuali-e-prospettive. Accessed 09 Dec 2020

Riahi S, Faridbod F, Ganjali MR (2009) Caffeine sensitive electrode and its analytical applications. Sens Lett 7(1):42–49. https://doi.org/10.1166/sl.2009.1008

Ristow M (2014) Unraveling the truth about antioxidants: mitohormesis explains ROS-induced health benefits. Nat Med 20(7):709–711. https://doi.org/10.1038/nm.3624

Sanchis-Gomar F, Pareja-Galeano H, Cervellin G, Lippi G, Earnest CP (2015) Energy drink over-consumption in adolescents: implications for arrhythmias and other cardiovascular events. Can J Cardiol 31(5):572–575. https://doi.org/10.1016/j.cjca.2014.12.019

Schinella G, Fantinelli JC, Mosca SM (2005) Cardioprotective effects of *Ilex paraguariensis* extract: evidence for a nitric oxide-dependent mechanism. Clin Nutr 24:360–366. https://doi.org/10.1016/j.clnu.2004.11.013

Schneider M (2017) Determinação da adulteração da erva-mate por adição de sacarose empregando espectroscopia no infravermelho (atr-ftir) em conjunto com ferramentas quimiométricas. Dissertation, Universidade Federal do Rio Grande do Sul, Porto Alegre

Senchina DS, Hallam JE, Kohut ML, Nguyen NA, Perera MADN (2014) Alkaloids and athlete immune function: caffeine, theophylline, gingerol, ephedrine, and their congeners. Exerc Immunol Rev 20:68–93

Singla RK, Dubey AK, Garg A, Sharma RK, Fiorino M, Ameen SM, Haddad MA, Al-Hiary M (2019) Natural polyphenols: chemical classification, definition of classes, subcategories, and structures. J AOAC Int 102(5):1397–1400. https://doi.org/10.1093/jaoac/102.5.1397

Tajik N, Tajik M, Mack I, Enck P (2017) The potential effects of chlorogenic acid, the main phenolic components in coffee, on health: a comprehensive review of the literature. Eur J Nutr 56(7):2215–2244. https://doi.org/10.1007/s00394-017-1379-1

Wang GX, Han J, Zhao LW, Jiang DX, Liu YT, Liu XL (2010) Anthelmintic activity of steroidal saponins from Paris polyphylla. Phytomedicine 17(14):1102–1105. https://doi.org/10.1016/j.phymed.2010.04.012

Chapter 5
Safety and Health Effects Ascribed to *Yerba Mate* Consumption

Abstract The connection between health and safety on the one hand and *Ilex paraguariensis* on the other side is basically dependent on the physical form of *yerba mate* products. *Yerba mate* beverages—obtained as infusions from dried and roasted or unroasted leaves—have been historically reported to have anti-oxidant power and different medical applications, also as ingredients for some folk medicine. All reported studies and also traditional histories concerning the 'miraculous' effects of *yerba mate* consumption on human health should be correlated with the peculiar mode of consumption: the 'infusion and consumption' cycle into *matero* containers. The regular assumption of *yerba mate* infusions in notable amounts, also depending on the hot or cold nature of infusion, can explain partially similar properties. Consequently, *chimarrão*, *tererê*, and *mate* teas have been studied with relation to new possible nutraceutical and functional foods/ingredients. With relation to safety and health features of *yerba mate* infusions, some of the most known key factors are briefly explained in this chapter.

Keywords Anti-inflammatory disease · Anti-oxidant power · Cancer · Cardiovascular disease · Lipid peroxidation · Oxidative stress · *Yerba mate*

5.1 Connections Between Chemical Profiles and Health Properties

The first part of this book has extensively discussed different matters, including *Yerba mate* health properties. Actually, the connection between health and safety on the one hand, and *I. paraguariensis* on the other side, is basically dependent on the physical form of *yerba mate* products. In fact, *I. paraguariensis* leaves are not consumed themselves; also, fruits and roots have no interests at present when speaking of potential foods or beverages with broad applications. On the other hand, *yerba mate* beverages—obtained as infusions from dried and roasted or unroasted leaves—have been historically reported to have anti-oxidant power and different medical applications, also as ingredients for some folk medicine.

It has been mentioned that bioactive compounds commonly found in *yerba mate*—polyphenols, saponins, xanthines, trace elements, and also vitamins—seem to be associated with positive features against several human diseases (Filip et al. 2010; Frizon et al. 2018; Heck and De Mejia 2007; Martins et al. 2009). In general, *yerba mate* may have some importance when speaking of human nutrition and supplementation against nutritional deficiencies. Naturally, all reported studies and also traditional histories concerning the 'miraculous' effects of *yerba mate* consumption on human health (with associated commercial claims) should be correlated with the peculiar mode of consumption (the 'infusion and consumption' cycle into *matero* containers). In fact, certain claimed properties should be critically examined and compared with real *yerba mate* features, including:

(a)   Regulation of the digestive apparatus
(b)   Skin protection
(c)   'Rejuvenating' power
(d)   Weight loss
(e)   Enhanced energy intake
(f)   Font of non-allergic caffeine (actually, the supplementation of 'mateine' is not different from caffeine: it is the same molecule!).

Certainly, the claimed presence of non-allergic mateine has to be highlighted… and the same thing when speaking of 'rejuvenating tea'…

The regular assumption of *yerba mate* infusions in notable amounts (500 ml at least per day), also depending on the hot or cold nature of infusion, can explain partially properties such as (Baeza et al. 2017; Cahuê et al. 2019; Cittadini et al. 2019a, b; Croge et al. 2020; Habtemariam 2019; Luís et al. 2019; Panza et al. 2018; Tate et al. 2020):

(1)   General anti-oxidant and anti-inflammatory features with associated hepato-protection and neuroprotection. Anti-oxidant power would be also… the cause of 'rejuvenation', with reference to skin protection
(2)   Hypoglycaemic power: glucose reduction and enhancement of sensitivity to insulin
(3)   Protective effects against hypertension, high cholesterol amounts in blood, cardiovascular diseases, oxidative stress in livers and hearth, post-ischaemic reperfusion and tissutal damages after heart attack, stroke and myocardial ischaemia, atherosclerosis, and cellular death
(4)   Protection against lipid peroxidation
(5)   Preventive effects against damages to the genetic human code
(6)   Synergic action with other therapies against cancer diseases
(7)   Regulation of the digestive apparatus
(8)   Anti-obesity power
(9)   Defence against oxidative stress (e.g. protection against retinal degeneration)
(10) Vasodilatator effects and other features related to the action of certain xanthines such as caffeine/matenine…

As a normal consequence, *yerba mate* has been and is still studied with relation to new possible nutraceutical and functional foods/ingredients. In particular, the possible addition of *yerba mate* extracts (*chimarrão, tererê*, and *mate* tea) to non-South American typical products or beverages such as soy beverages (Frizon et al. 2018) has been proposed. In fact, this plant has been often considered as a 'carrier' beverage for characteristic medicinal plants, while *I. paraguariensis* is not historically considered a medicinal plant. The fact that Paraguayans and Argentineans consider *yerba mate* also as a medicinal remedy has to be highlighted. A good explanation is that *chimarrão, tererê*, and *mate* tea infusions can show a synergetic function in association with other vegetable remedies (Arenas and Azorero 1977; Goyke 2017; Kujawska 2018; Kujawska et al. 2017; Schmeda-Hirschmann and Bordas 1990). On the other side, it has been reported in recent years that *yerba mate* consumption could be a risk factor when speaking of oral tumours.

Finally, it has been sometimes reported that the notable *yerba mate* consumption could be linked with increased risk of various tumours (Bates et al. 2007; Frizon et al. 2018; Goldenberg et al. 2003; Heck and De Mejia 2007). It has been suggested that a possible association exists when tobacco and/or alcohol act synergically with *yerba mate* consumption. More research is still needed in this ambit (Aguilera 2015; Deneo-Pellegrini et al. 2013; Stefani et al. 2011). With relation to safety and health features of *yerba mate*, some of the most known factors are briefly explained in the next sections.

5.2 General Anti-oxidant and Anti-inflammatory Features—Defence Against Oxidative Stress. Defence Against Cardiovascular Diseases

Anti-oxidant and anti-inflammatory properties are generally ascribed to polyphenols of vegetable origin, with concern to *yerba mate* and other plants. In this ambit, polyphenols are certainly one of the most promising natural substances with similar features (Barbera 2020; Barbieri et al. 2019; Bhagat et al. 2019; Delgado et al. 2016, 2019; Haddad et al. 2020a, b; Issaoui et al. 2020; Matta 2019; Parisi 2019, 2020; Parisi and Dongo 2020; Parisi et al. 2020; Singla et al. 2019) because of their peculiar chemical structure and their function in oxidation and reduction reactions. For these reasons, various properties ascribed to *yerba mate* ('rejuvenation'… or skin protection, defence against ageing in human beings, protection against retinal degeneration, etc.) are justified. In addition, the following features (normally important in plants) should be considered in the human being:

(1) Colour enhancement
(2) Defence against ultraviolet rays
(3) Defence against toxic heavy metals by means of chelation effect.

In this ambit, polyphenols are able to enhance the performance of enzymatic systems and reactions relying on oxidation–reduction reactions by means of peroxidase-like activity (it can block chemically reactive oxygen species). All possible cardiovascular diseases and other illnesses can be contrasted with the action of phenolics, including the powerful chlorogenic acid (and its isomers). Interestingly, the effect of *mate* consumption appears more pronounced in the human being if compared with the consumption of red wine (with concern to the 'French Paradox') and green tea.

5.3 Protection Against Diabetes, Cancer, and Lipid Peroxidation

As mentioned above, polyphenols have a plethora of anti-oxidant features related to oxidation and reduction reactions. The claimed reduction of lipid peroxidation is ascribed to polyphenols, including the powerful chlorogenic acid.

The protective action of polyphenols includes also useful effects against type-2 diabetes. A possible explanation may be partially linked with the relationship between *yerba mate* extracts and the dose-dependent inhibition of dicarbonyl action (dicarbonyls are involved in the production of advanced glycation end products, while oxidation is correlated to glycation) (Heck and De Mejia 2007).

With reference to cancer treatments, *yerba mate* has been reported to have interesting inhibitory properties against several mechanisms involved in tumours. Actually, the efficacy of these properties seems to be dependent on the concomitant presence of different molecules, including chlorophyll and ursolic acid (group of triterpenoid saponins) (Heck and De Mejia 2007).

Finally, the reductions of lipid peroxidation, and also the reduction of absorption of cytotoxic lipid peroxidation products, have been always reported when speaking of polyphenols in red wines, as an attempt to clarify protection mechanisms in the 'French Paradox' and the usefulness of Japanese and Mediterranean diet styles (Gorelik et al. 2008). *Yerba mate* effects can be partially explained in this way.

References

Aguilera JM (2015) *Ilex paraguariensis* (yerba mate) infusions and risk of oral cancer: a structured literature review. University of Ottawa, Ottawa. Available https://ruor.uottawa.ca/bitstream/10393/33451/1/Yerba%20Mate%20and%20Oral%20Cancer.pdf. Accessed 10 Dec 2020

Arenas P, Azorero R (1977) Plants of common use in Paraguayan folk medicine for regulating fertility. Econ Bot 31(3):298–300. https://doi.org/10.1007/BF02866879

Baeza G, Sarriá B, Bravo L, Mateos R (2017) Polyphenol content, in vitro bioaccessibility and antioxidant capacity of widely consumed beverages. J Sci Food Agric 98(4):1397–1406. https://doi.org/10.1002/jsfa.8607

Barbera M (2020) Reuse of food waste and wastewater as a source of polyphenolic compounds to use as food additives. J AOAC Int 103(4):906–914. https://doi.org/10.1093/jaocint/qsz025

Barbieri G, Bergamaschi M, Saccani G, Caruso G, Santangelo A, Tulumello T, Vibhute B, Barbieri G (2019) Processed meat and polyphenols: opportunities, advantages, and difficulties. J AOAC Int 102(5):1401–1406. https://doi.org/10.1093/jaoac/102.5.1401

Bates MN, Hopenhayn C, Rey OA, Moore LE (2007) Bladder cancer and mate consumption in Argentina: a case-control study. Cancer Lett 246(1–2):268–273. https://doi.org/10.1016/j.canlet. 2006.03.005

Bhagat AR, Delgado AM, Issaoui M, Chammem N, Fiorino M, Pellerito A, Natalello S (2019) Review of the role of fluid dairy in delivery of polyphenolic compounds in the diet: chocolate milk, coffee beverages, Matcha green tea, and beyond. J AOAC Int 102(5):1365–1372. https://doi.org/10.1093/jaoac/102.5.1365

Cahuê F, Nascimento JHM, Barcellos L, Salerno VP (2019) *Ilex paraguariensis*, exercise and cardioprotection: a retrospective analysis. J Funct Foods 53:105–108. https://doi.org/10.1016/j.jff.2018.12.008

Cittadini MC, Albrecht C, Miranda AR, Mazzuduli GM, Soria EA, Repossi G (2019a) Neuroprotective effect of *Ilex paraguariensis* intake on brain myelin of lung adenocarcinoma-bearing male Balb/c mice. Nutr Cancer 71(4):629–633. https://doi.org/10.1080/01635581.2018.1559932

Cittadini MC, Repossi G, Albrecht C, Di Paola Naranjo R, Miranda AR, Pascual-Teresa S, Soria EA (2019b) Effects of bioavailable phenolic com-pounds from *Ilex paraguariensis* on the brain of mice with lung adenocarcinoma. Phytother Res 33(4):1142–1149. https://doi.org/10.1002/ptr. 6308

Croge CP, Cuquel FL, Pintro PTM (2020) Yerba mate: cultivation systems, processing and chemical composition. A review. Sci Agric 78(5):e20190259. https://doi.org/10.1590/1678-992X-2019-0259

Delgado AM, Vaz de Almeida MD, Parisi S (2016) Chemistry of the Mediterranean diet. Springer International Publishing, Cham. https://doi.org/10.1007/978-3-319-29370-7

Delgado AM, Issaoui M, Chammem N (2019) Analysis of main and healthy phenolic compounds in foods. J AOAC Int 102(5):1356–1364. https://doi.org/10.1093/jaoac/102.5.1356

Deneo-Pellegrini H, De Stefani E, Boffetta P, Ronco AL, Acosta G, Correa P, Mendilaharsu M (2013) Maté consumption and risk of oral cancer: case-control study in Uruguay. Head Neck 35(8):1091–1095. https://doi.org/10.1002/hed.23080

Filip R, Davicino R, Anesini C (2010) Antifungal activity of the aqueous extract of *Ilex paraguariensis* against *Malassezia furfur*. Phytother Res 24(5):715–719. https://doi.org/10.1002/ptr. 3004

Frizon CNT, Perussello CA, Sturion JA, Hoffmann-Ribani R (2018) Novel beverages of yerba-mate and soy: bioactive compounds and functional properties. Beverages 4(1):21. https://doi.org/10. 3390/beverages4010021

Goldenberg D, Golz A, Joachims HZ (2003) The beverage maté: a risk factor for cancer of the head and neck. Head Neck 25(7):595–601. https://doi.org/10.1002/hed.10288

Gorelik S, Ligumsky M, Kohen R, Kanner J (2008) A novel function of red wine polyphenols in humans: prevention of absorption of cytotoxic lipid peroxidation products. FASEB J 22(1):41–46. https://doi.org/10.1096/fj.07-9041com

Goyke N (2017) Traditional medicinal use in Chamorro Cué, Gral. E. Aquino, San Pedro, Paraguay. Dissertation, Michigan Technological University, Houghton

Habtemariam S (2019) The chemical and pharmacological basis of yerba maté (*Ilex paraguariensis* A. St.-Hil.) as potential therapy for type 2 diabetes and metabolic syndrome. In: Habtemariam S (ed) Medicinal foods as potential therapies for type-2 diabetes and associated diseases. Academic Press, New York, pp 943–983

Haddad MA, Dmour H, Al-Khazaleh JFM, Obeidat M, Al-Abbadi A, Al-Shadaideh AN, Al-Mazra'awi MS, Shatnawi MA, Iommi C (2020a) Herbs and medicinal plants in Jordan. J AOAC Int 103(4):925–929. https://doi.org/10.1093/jaocint/qsz026

Haddad MA, El-Qudah J, Abu-Romman S, Obeidat M, Iommi C, Jaradat DSM (2020b) Phenolics in Mediterranean and Middle East important fruits. J AOAC Int 103(4):930–934. https://doi.org/10.1093/jaocint/qsz027

Heck CI, De Mejia EG (2007) Yerba mate tea (*Ilex paraguariensis*): a comprehensive review on chemistry, health implications, and technological considerations. J Food Sci 72(9):R138–R151. https://doi.org/10.1111/j.1750-3841.2007.00535.x

Issaoui M, Delgado AM, Caruso G, Micali M, Barbera M, Atrous H, Ouslati A, Chammem N (2020) Phenols, flavors, and the Mediterranean diet. J AOAC Int 103(4):915–992. https://doi.org/10.1093/jaocint/qsz018

Kujawska M (2018) Yerba mate (*Ilex paraguariensis*) beverage: nutraceutical ingredient or conveyor for the intake of medicinal plants? Evidence from Paraguayan folk medicine. Evid Based Complement Alternat Med 2018:6849317. https://doi.org/10.1155/2018/6849317

Kujawska M, Jiménez-Escobar MND, Nolan JM, Arias-Mutis D (2017) Cognition, culture and utility: plant classification by Paraguayan immigrant farmers in Misiones, Argentina. J Ethnobiol Ethnomed 13(1):42. https://doi.org/10.1186/s13002-017-0169-4

Luís AFS, Domingues FC, Pereira Amaral LMJ (2019) The anti-obesity potential of *Ilex paraguariensis*: results from a meta-analysis. Braz J Pharm Sci 55:e17615. https://doi.org/10.1590/s2175-97902019000217615

Martins F, Suzan AJ, Cerutti SM, Arçari DP, Ribeiro ML, Bastos DH, Carvalho Pde O (2009) Consumption of mate tea (*Ilex paraguariensis*) decreases the oxidation of unsaturated fatty acids in mouse liver. Br J Nutr 101(4):527–532. https://doi.org/10.1017/S000711450802504X

Matta FV (2019) Chemical analysis of typical beverages and Açaí Berry from South America. Dissertation, University of Surrey, Guildford

Panza VP, Brunetta HS, Oliveira MV, Nunes EA, Silva EL (2018) Effect of mate tea (*Ilex paraguariensis*) on the expression of the leukocyte NADPH oxidase subunit p47^{phox} and on circulating inflammatory cytokines in healthy men: a pilot study. Int J Food Sci Nutr 70:212–221. https://doi.org/10.1080/09637486.2018.1486393

Parisi S (2019) Analysis of major phenolic compounds in foods and their health effects. AOAC J 102(5):1354–1355. https://doi.org/10.5740/jaoacint.19-0127

Parisi S (2020) Characterization of major phenolic compounds in selected foods by the technological and health promotion viewpoints. J AOAC Int 103(4):904–905. https://doi.org/10.1093/jaoacint/qsaa011

Parisi S, Dongo D (2020). Polifenoli e salute. I vegetali amici del sistema immunitario. Great Italian Food Trade. Available https://www.greatitalianfoodtrade.it/salute/polifenoli-e-salute-i-vegetali-amici-del-sistema-immunitario. Accessed 09 Dec 2020

Parisi S, Dongo D, Parisi C (2020) Resveratrolo, conoscenze attuali e prospettive. Great Italian Food Trade. Available www.greatitalianfoodtrade.it/integratori/resveratrolo-conoscenze-attuali-e-prospettive. Accessed 09 Dec 2020

Schmeda-Hirschmann G, Bordas E (1990) Paraguayan medicinal compositae. J Ethnopharmacolvol 28(2):163–171. https://doi.org/10.1016/0378-8741(90)90026-P

Singla RK, Dubey AK, Garg A, Sharma RK, Fiorino M, Ameen SM, Haddad MA, Al-Hiary M (2019) Natural polyphenols: chemical classification, definition of classes, subcategories, and structures. J AOAC Int 102(5):1397–1400. https://doi.org/10.1093/jaoac/102.5.1397

Stefani ED, Moore M, Aune D, Deneo-Pellegrini H, Ronco AL, Boffetta P, Correa P, Acosta G, Mendilaharsu M, Luaces ME, Silva C, Landó G (2011) Maté consumption and risk of cancer: a multi-site case-control study in Uruguay. Asian Pac J Cancer Prev 12(4):1089–1093

Tate PS, Marazita MC, Marquioni-Ramella MD, Suburo AM (2020) *Ilex paraguariensis* extracts and its polyphenols prevent oxidative damage and senescence of human retinal pigment epithelium cells. J Funct Food 67:103833. https://doi.org/10.1016/j.jff.2020.103833

Chapter 6
Yerba Mate, Tereré or *Chimarrão*? Different Countries, Historical Legacy, and Similar Preparations

Abstract This chapter concerns the cultural history of *Ilex paraguariensis* as a single vegetable species belonging to the *Aquifoliaceae* family, found naturally in Brazil, Argentina, Paraguay, and Uruguay. The original name recalls immediately the Paraguay name because of its original habitat, the humid forests surrounding the Paraná River in eastern Paraguay. Historically, the *yerba mate* infusion is known for centuries, before the arrival of Spanish *conquistadores* and the Jesuits. After the expulsion of the Jesuits, *yerba mate* remained a cultural tradition for the *Mestizo* population, the local *criollo*s. Nowadays, *yerba mate* is extremely popular in South America and other continents. However, there are different *yerba mate* products, depending on processes, packaging choices, national or regional habits, and also the mode of consumption. The non-accustomed consumer is probably unaware of this situation. The cultural heritage of *yerba mate* is a notable key factor explaining partially the success of this plant worldwide.

Keywords Chimarrão · Cultural heritage · Guaranì · *Ilex paraguariensis* · Mestizo · Tererê · *Yerba mate*

6.1 *Yerba Mate*, History, and National/Regional Traditions

Ilex paraguariensis is a single species belonging to the *Aquifoliaceae* family, and it is found naturally in the following South American countries: Brazil, Argentina, Paraguay, and Uruguay. The original name—*I. paraguariensis*—recalls immediately the Paraguay name because of its original habitat, the humid forests surrounding the Paraná River in eastern Paraguay (Folch 2010). Historically, the *yerba mate* infusion is known for centuries, before the arrival of Spanish *conquistadores* and the Jesuits. In fact, the ancient tradition is a cultural heritage of the *Guaraní* population. Because of the commercial interest of the Jesuits, *yerba mate*—*Guaraní* name: *ka'a*; commercial name: 'Jesuit tea' or 'Paraguayan tea'—became extremely popular in South America, with some export activity to Europe. After the expulsion of the Jesuits, *yerba mate* remained a cultural tradition for the *Mestizo* population (Spanish and *Guaraní* mixed origin), the local *criollo*s (Kujawska 2018; Rau 2009).

© The Author(s), under exclusive license to Springer Nature Switzerland AG 2021
C. Iommi, *Chemistry and Safety of South American Yerba Mate Teas*,
Chemistry of Foods, https://doi.org/10.1007/978-3-030-69614-6_6

At present, *yerba mate* is extremely popular in South America and other continents. However, there are different *yerba mate* products, depending on processes, packaging choices, national or regional habits, and also the mode of consumption. The non-accustomed consumer is probably unaware of this situation. As a result, this chapter would discuss historical traditions linked with *yerba mate* (including different names for the same product, depending on the nationality!) and the peculiarity of three different *yerba mate* uses. The cultural heritage of *yerba mate* is a notable key factor explaining partially the success of this plant worldwide. It has to be remembered that cultural traditions and peculiar features of *yerba mate* preparations may also differ depending on claimed nutritional or health properties, social diffusion, reliable traceability information at present and in the past, packaging design and choices, chemical composition in terms of phenolic compounds, xanthines, saponins, etc. (Barbieri et al. 2014, 2019; Bhagat et al. 2019; Brunazzi et al. 2014; Chammem et al. 2018; Delgado et al. 2016a, b, 2017, 2019; Fiorino et al. 2019; Haddad et al. 2020a, b; Giberti 1995; Issaoui et al. 2020; Loria et al. 2009; Messina et al. 2015; Mania et al. 2016a, b, 2018; Parisi 2016, 2019; Parisi et al. 2020; Piovezan-Borges et al. 2016).

6.2 *Yerba Mate* Types

From the commercial viewpoint, eight different qualities of *yerba mate* can be found on the market (Matta 2019):

(a) *Premium*
(b) *Nativa*
(c) *Traditional,*
(d) *Moida grossa*
(e) *Exportação*
(f) *Tererê*
(g) *Tostada*
(h) *Cambona.*

The chemical and technological differences linked to these eight *yerba mate* products have been discussed in Sect. 1.4.1 (Matta 2019). A different classification is related to the packaging type (influencing the final use):

(1) Teabag-packaged *yerba mate*
(2) Loose-packaged *yerba mate.*

In addition, the colour of *yerba mate* leaves has its own causes and importance: green or roasted (brownish) leaves. As a result, the following variables have to be taken into account (Sect. 1.4.1) (Matta 2019):

(a)   The selection of leaves
(b)   The choice between traditional or plantation method
(c)   The choice of triturating and drying process (*barbaqua*/*secado* procedure)
(d)   The selection of qualities according to particle sizes and additional ageing periods
(e)   The possible reduction of twigs
(f)   The presence or absence of the roasting process
(g)   The use of genetic variations of *I. paraguariensis.*

Anyway, the eight *yerba mate* qualities are used for infusion purposes (Bastos et al. 2007; Heck and De Mejia 2007; Matta 2019):

(1)   Hot (also named 'regular') infusions in water. The correspondent names are *chimarrão* (from green and dried leaves) or *mate* (from roasted/brownish leaves)
(2)   Cold infusions in water. The correspondent name is generally *tererê* (from green and dried leaves).

6.3   National Differences… Why?

National preferences have some importance when speaking of *chimarrão, mate,* and *tererê* infusions.

In general, the *mate* infusion seems to be preferred in Argentina and Uruguay, while *chimarrão* appears to be preferred by Brazilian and Uruguayan consumers. Finally, *tererê* is extremely appreciated in Paraguay, and *mate* is also preferred in this country (Folch 2010; Kujawska 2018).

The apparent difference between national preferences for *yerba mate* consumption could be studied on the basis of the historical *Guaraní* and *Mestizo* traditions (Delacassa and Bandoni 2001; Kujawska 2018). The type of consumption cannot be separated or fully understood without the peculiar tool for use (the traditional gourd-shaped container, named *mate, matero, porongo matero or yerbamatero*) (Folch 2010; Oberti 1960).

However, and based on the existing evidences, it appears that the main difference does not concern ethnic or cultural traditions themselves as the result of environment and social behaviours. On the contrary, the difference between hot/regular (*mate, chimarrão*) and cold (*tererê*) infusions is simply linked to seasons and weather conditions. In summary, hot/regular infusions are preferred all year, with a specific preference for morning and afternoon social habits. For this reason, *mate* and *chimarrão* are named 'regular' infusions. On the other side, cold (*tererê*) infusions are essentially and naturally preferred in hot days (the summer season: from December to March), with preference for afternoon occasions, and also before lunch (Kujawska 2018).[1]

---

[1]https://www.hindawi.com/journals/ecam/2018/6849317/.

Finally, it should be remembered (Chap. 1) that the main producers of *yerba mate* are Argentina and Brazil. The third country, Uruguay, usually imports *yerba mate* leaves from Brazil or Argentina, finishing the preparation of *yerba mate* leaves locally. This clarification has to be taken into account because the use of *chimarrão* type in Uruguay is extremely diffused (although the raw material is substantially imported from other countries), becoming one of the most important social practices in this country.

References

Barbieri G, Barone C, Bhagat A, Caruso G, Conley Z, Parisi S (2014) The prediction of shelf life values in function of the chemical composition in soft cheeses. In: The Influence of chemistry on new foods and traditional products. Springer International Publishing, Cham. https://doi.org/10.1007/978-3-319-11358-6_2

Barbieri G, Bergamaschi M, Saccani G, Caruso G, Santangelo A, Tulumello T, Vibhute B, Barbieri G (2019) Processed Meat and Polyphenols: Opportunities, Advantages, and Difficulties. J AOAC Int 102(5):1401–1406. https://doi.org/10.1093/jaoac/102.5.1401

Bastos DHM, De Oliveira DM, Matsumoto RT, Carvalho PDO, Ribeiro ML (2007) Yerba mate: pharmacological properties, research and biotechnology. Med Aromat Plant Sci Biotechnol 1(1):37–46

Bhagat AR, Delgado AM, Issaoui M, Chammem N, Fiorino M, Pellerito A, Natalello S (2019) Review of the role of fluid dairy in delivery of polyphenolic compounds in the diet: chocolate milk, coffee beverages, Matcha green tea, and beyond. J AOAC Int 102(5):1365–1372. https://doi.org/10.1093/jaoac/102.5.1365

Brunazzi G, Parisi S, Pereno A (2014) The importance of packaging design for the chemistry of food products. Springer International Publishing, Cham. https://doi.org/10.1007/978-3-319-084 52-7

Chammem N, Issaoui M, De Almeida AID, Delgado AM (2018) Food crises and food safety incidents in European Union, United States, and Maghreb Area: current risk communication strategies and new approaches. J AOAC Int 101(4):923–938. https://doi.org/10.5740/jaoacint.17-0446

Delacassa E, Bandoni AL (2001) El mate. Revista de Fitoterapia 1(4):257–265

Delgado AM, Almeida MDV, Parisi S (2017) Chemistry of the mediterranean diet. Springer International Publishing, Cham. https://doi.org/10.1007/978-3-319-29370-7

Delgado AM, Issaoui M, Chammem N (2019) Analysis of main and healthy phenolic compounds in foods. J AOAC Int 102(5):1356–1364. https://doi.org/10.1093/jaoac/102.5.1356

Delgado AM, Vaz de Almeida MD, Barone C, Parisi S (2016a) Leguminosas na Dieta Mediterrânica—Nutrição, Segurança, Sustentabilidade. CISA—VIII Conferência de Inovação e Segurança Alimentar, ESTM- IPLeiria, Peniche, Portugal

Delgado AM, Vaz de Almeida MD, Parisi S (2016b) Chemistry of the mediterranean diet. Springer International Publishing, Cham. https://doi.org/10.1007/978-3-319-29370-7

Fiorino M, Barone C, Barone M, Mason M, Bhagat A (2019) The intentional adulteration in foods and quality management systems: chemical aspects. Qual Syst Food Indus. Springer International Publishing, Cham, pp 29–37

Folch C (2010) Stimulating consumption: yerba mate myths, markets, and meanings from conquest to present. Comp Stud Soc Hist 52(1):6–36. https://doi.org/10.1017/S0010417509990314

Giberti GC (1995) Aspectos oscuros de la corología de Ilex paraguariensis St. Hil. In: Winge H, Ferreira AG, de Araujo Mariath JE, Tarasconi LC (eds) "Erva-Mate: biologia e cultura no Cone Sul, pp. 289–300. Editora da UFRGS, Porto Alegre

Haddad MA, Dmour H, Al-Khazaleh JFM, Obeidat M, Al-Abbadi A, Al-Shadaideh AN, Al-mazra'awi MS, Shatnawi MA, Iommi C (2020a) Herbs and medicinal plants in Jordan. J AOAC Int 103(4):925–929. https://doi.org/10.1093/jaocint/qsz026

Haddad MA, El-Qudah J, Abu-Romman S, Obeidat M, Iommi C, Jaradat DSM (2020b) Phenolics in mediterranean and middle east important fruits. J AOAC Int 103(4):930–934. https://doi.org/10.1093/jaocint/qsz027

Heck CI, De Mejia EG (2007) Yerba Mate Tea (Ilex paraguariensis): a comprehensive review on chemistry, health implications, and technological considerations. J Food Sci 72(9):R138–R151. https://doi.org/10.1111/j.1750-3841.2007.00535.x

Issaoui M, Delgado AM, Caruso G, Micali M, Barbera M, Atrous H, Ouslati A, Chammem N (2020) Phenols, flavors, and the mediterranean diet. J AOAC Int 103(4):915–992. https://doi.org/10.1093/jaocint/qsz018

Kujawska M (2018) Yerba mate (Ilex paraguariensis) beverage: nutraceutical ingredient or conveyor for the intake of medicinal plants? Evidence from Paraguayan folk medicine. Evid Based Compl Alt 2018:6849317. https://doi.org/10.1155/2018/6849317

Loria D, Barrios E, Zanetti R (2009) Cancer and yerba mate consumption: a review of possible associations. Rev Panam Salud Púb 25:530–539

Mania I, Barone C, Caruso G, Delgado A, Micali M, Parisi S (2016a) Traceability in the cheese-making field. The regulatory ambit and practical solutions. Food Qual Mag 3:18–20. ISSN 2336–4602

Mania I, Delgado AM, Barone C, Parisi S (2018) Traceability in the dairy industry in Europe. Springer International Publishing, Heidelberg, Germany

Mania I, Fiorino M, Barone C, Barone M, Parisi S (2016b) Traceability of packaging materials in the cheesemaking field. The EU Regulatory Ambit. Food Packag Bull 25, 4&5:11–16

Matta FV (2019) Chemical analysis of typical beverages and açaí berry from south america dissertation. University of Surrey, Guildford

Messina D, Soto C, Méndez A, Corte C, Kemnitz M, Avena V, Dal Balzo D, Pérez Elizalde R (2015) Efecto hipolipemiante del consumo de mate en individuos dislipidémicos. Nutr Hosp 31(5):2131–2139

Oberti F (1960) Disquisiciones sobre el origen de la bombilla. Cuadernos del Instituto Nacional de Antropología y Pensamiento Latinoamericano 1:151–158

Parisi S (2016) The world of foods and beverages today: globalization, crisis management and future perspectives. learning.ly/ The Economist Group, available http://learning.ly/products/the-world-of-foods-and-beverages-today-globalization-crisis-management-and-future-perspectives. Accessed 9 Dec 2020

Parisi S (2019) Analysis of major phenolic compounds in foods and their health effects. AOAC J 102(5):1354–1355. https://doi.org/10.5740/jaoacint.19-0127

Parisi S, Dongo D, Parisi C (2020) Resveratrolo, conoscenze attuali e prospettive. Great Italian Food Trade 27/10/2020. Available www.greatitalianfoodtrade.it/integratori/resveratrolo-conosc enze-attuali-e-prospettive. Accessed 9 Dec 2020

Piovezan-Borges AC, Valério-Júnior C, Gonçalves IL, Mielniczki-Pereira AA, Valduga AT (2016) Antioxidant potential of yerba mate (Ilex paraguariensis St. Hil.) extracts in Saccharomyces cerevisae deficient in oxidant defense genes. Braz J Biol 76, 2:539–544. https://doi.org/10.1590/1519-6984.01115

Rau V (2009) La yerba mate en misiones (Argentina). Estructura y significados de una producción localizada. Rev Agroaliment 15, 28:49–58

Chapter 7
Regulatory Aspects Concerning *Yerba Mate* Tea

Abstract The regulatory situation concerning *yerba mate* commerce in South America and worldwide should be discussed. *Yerba mate* is a critical economic resource for South American people working in farms and associated industries. As a single example, this plant has been defined the national infusion in Argentina some year ago, and similar feelings can be easily observed in Brazil and other South American countries (Brazil, Uruguay, and Paraguay) also meaning that all included economies rely on this product with a peculiar importance for several regions only. At present, yerba mate represents a global commerce worldwide with notable economic incomes. However, the ineffective control and management of the agribusiness industry have forced many small farmers to abandon their homes and cultivations because of the impossibility to guarantee high production amounts. *Yerba mate* market is essentially an oligopolistic system, and the same thing is true for the past. The commercialisation of fair trade and organic *yerba mate* products has partially determined some amelioration for agricultural workers because of the interest in ethical policies and ethnic products (alternative foods). As a result, more regulation should be needed. This chapter concerns the regulatory situation in Argentina and Brazil, the two main *yerba mate* producers.

Keywords Authenticity · European union · Geographical indication · INYM · MERCOSUR · South america · *Yerba mate*

Abbreviations

DOU	Diário Oficial da União
EU	European Union
GI	Geographical indication
INYM	*Instituto Nacional de la Yerba Mate*
MERCOSUR	*Mercado Común del Sur*

7.1   The Global Importance of *Yerba Mate*

Yerba mate is a critical economic resource for South American people working in farms and associated industries. The example of Argentina is notable: *yerba mate* has been defined the national infusion in this country in 2013, and similar feelings can be easily observed in remaining countries (Gortari 2007; Lawson 2009; Smith 2014), although it has to be considered that Uruguay imports leaves with the final production on site. On the other side, Brazil, Argentina, and Paraguay have the complete '*yerba mate* supply chain'. Anyway, the '*yerba mate*' region—boundaries: 21°00'00" S; 30°00'00" S (latitudes); 48°30'00" W, 56°10'00" W (longitudes) (Croge et al. 2020)—is common to all the above-mentioned countries, also meaning that all included economies rely on this product with a peculiar importance for several regions only.

At present, *yerba mate* represents a global commerce worldwide with notable economic incomes. However, the ineffective control and management of the agribusiness industry have forced many small farmers to abandon their homes and cultivations because of the impossibility to guarantee high production amounts. *Yerba mate* market is essentially an oligopolistic system, and the same thing is true for the past. The commercialisation of ethic (fair trade) and organic *yerba mate* products has partially determined some amelioration for agricultural workers (Smith 2014) because of the interest in ethical policies and ethnic products (alternative foods). As a result, more regulation should be needed. The next section gives a brief example of the regulatory situation in Argentina and Brazil (the two main *yerba mate* producers).

7.2   *Yerba Mate* Regulation in Argentina and Brazil

The commercial development and basic rules for *yerba mate* productions are managed, in Argentina, in the ambit of the *Instituto Nacional de la Yerba Mate* (INYM), ruling the whole sector and also carrying on the general promotion of *yerba mate* in Argentina and worldwide (Lawson 2009; Ministerio de Agroindustria 2016). However, it has to be considered that the oligopolistic nature of the *yerba mate* system in Argentina is still the normal status. Consequently, sellers notably surpass the number of potential buyers, and this disequilibrium needs more regulatory norms able to defend the Argentinean economic system and also *yerba mate* consumers, for a commerce which allows nowadays selling and buying this product worldwide.

From the regulatory viewpoint, the Argentinean INYM has finally obtained the geographical indication (GI) for '*yerba mate*' in 2016 (Moeller IP Advisors 2016). Thus, the INYM is now completely able to guarantee—and defend—quality distinctions (Sect. 1.4), legal protection, and information related to the origin of *yerba mate*. In this country, the main regulatory document is the *Ley N° 18.284—Código Alimentario Argentino* (Law N° 18.284—Argentinean Food Code), established in 1969. This law introduces technical norms with concern to hygiene-, safety-, and commercially oriented provisions related to *yerba mate* commerce (Poder Ejecutivo Nacional 1969). The last amendment in force is actually the *Resolución Conjunta Nº 43/2013 y 59/2013—Modificación del Código Alimentario Argentino* (1 February 2013) (Secretaría de Políticas, Regulación e Institutos, and Secretaría de Agricultura, Ganadería y Pesca 2013). The interested reader is invited to consult the main regulation and its amendments. Another interesting document, in relation to *yerba mate* commerce, is the Resolución Nº 9/2017—Apruébase el Régimen de Documentación de Movimientos de *Yerba Mate* (19 January 2017) (INYM 2017), lastly modified by Resolution 153/2017.

With reference to Brazil, the second *yerba mate* producer, a 'National Policy on *Yerba Mate*', has been issued in Brazil (3 January 2019) with the aim of enhancing the production, quality, and the general commerce for this vegetable product (Presidência da República 2019). The document has the same objectives of the Argentinean INYM, including not only the safety of consumers, but also the defence of environmental sustainability.

For the rest, regulatory norms in South America often refer to particular documents of the *Mercado Común del Sur* (MERCOSUR) (European Commission 2019), the common South American market, and agreements with other economic areas such as the European Union (EU) (European Commission 2020). According to the Trade Part of the Agreement following the agreement in principle announced on 28 June 2019 (Agreement between the EU and the MERCOSUR) (European Commission 2019), *yerba mate* is considered in the category of 'beverages' when speaking of '*Yerba Mate Argentina/ Yerba Mate Elaborada con Palo*' (origin: Argentina), 'other beverages' as '*Yerbamate Paraguaya*' (Paraguay), or '*mate* herb' with concern to '*São Matheus*' (Brazil). General features concerning *yerba mate* are discussed in the MERCOSUR's Technical Regulation for the Labelling of Packed Food, Resolution Nº 26 /2003 (Sniechowski and Paul 2008), influencing also National Regulations such as the Uruguay's *Decreto Nº 32/015—Límites de contaminantes en yerba mate*, with reference to limits of contaminant in this product. The interested reader is invited to consult the related literature.

References

Croge CP, Cuquel FL, Pintro PTM (2020) Yerba mate: cultivation systems, processing and chemical composition. A review Scientia Agricola 78(5):e20190259. https://doi.org/10.1590/1678-992X-2019-0259

European Commission (2019) New EU-Mercosur trade agreement—the agreement in principle, Brussels, 1 July 2019. European Commission, Brussels. Available https://trade.ec.europa.eu/doc lib/docs/2019/june/tradoc_157964.pdf. Accessed 11 Dec 2020

European Commission (2020) EU Novel food catalogue. European Commission, Brussels. Available https://ec.europa.eu/food/safety/novel_food/catalogue/search/public. Accessed 11 Dec 2020

Gortari J (2007) El Instituto Nacional de la Yerba Mate (INYM) como dispositivo político de economía social". In: Gortari J (ed) De la tierra sin mal al tractorazo. Hacia una economía política de la yerba mate. Ed. Universitaria, Universidad Nacional de Misiones, Posadas

INYM (2017) Resolución 9/2017—Apruébase el Régimen de Documentación de Movimientos de Yerba Mate. Boletín Oficial N° 33.551, 24 de enero de 2017:16. Available http://faolex.fao.org/docs/pdf/arg162346.pdf. Accessed 15 Dec 2020

Lawson J (2009) Cultivating green gold: a political ecology of land use changes for small yerba mate farmers in misiones, Argentina. Dissertation, Yale School of Forestry, New Haven, CT

Ministerio de Agroindustria (2016) Resolución N° 13/2016—Aprueba el Protocolo de Producción, Elaboración y/o Guía de Prácticas de la 'Yerba Mate Argentina'. Boletín Oficial N° 33337, 15 de marzo de 2016, p. 31. Available http://www.fao.org/faolex/results/details/en/c/LEX-FAOC153569. Accessed 15 Dec 2020

Moeller IP Advisors (2016) Argentina: yerba mate is a new GI in Argentina. Mondaq® Ltd, London, New York, and Sidney. Available https://www.mondaq.com/argentina/trademark/552298/yerba-mate-is. Accessed 11 Dec 2020

Poder Ejecutivo Nacional (1969) Ley N° 18.284—Código Alimentario Argentino. Boletín Oficial N° 21.732, 28 de julio de 1969:1–2. Available http://www.fao.org/faolex/results/details/en/c/LEX-FAOC011896. Accessed 15 Dec 2020

Presidência da República (2019) Lei n. 13.791—Dispõe sobre a Política Nacional da Erva-Mate. Presidência da República, Casa Civil, Subchefia para Assuntos Jurídicos, Brazil. Diário Oficial da União (DOU) of 04 January 2019. Available http://www.fao.org/faolex/results/details/en/c/LEX-FAOC183890/. Accessed 11 Dec 2020

Secretaría de Políticas, Regulación e Institutos, and Secretaría de Agricultura, Ganadería y Pesca (2013) Resolución Conjunta N° 43/2013 y 59/2013—Modificación del Código Alimentario Argentino. Boletín Oficial N° 32.574, 1° de febrero de 2013:22. Available http://www.fao.org/faolex/results/details/en/c/LEX-FAOC119910. Accessed 15 Dec 2020

Smith F (2014) Exploring fair trade yerba mate networks in misiones, Argentina. Dissertation, University of Miami, Miami. Available https://scholarship.miami.edu/discovery/fulldisplay/alma991031447938602976/01UOML_INST:ResearchRepository?tags=scholar. Accessed 10 Dec 2020

Sniechowski VI, Paul LM (2008) The labeling on the" yerba mate"(Ilex Paraguariensis) packages in the Mercosur"(South American Common Market). Visión de Futuro 9(1):105–124

Chapter 8
Yerba Mate, the Global Commerce, and Possible Adulteration. The Current Situation and New Perspectives

Abstract Commercial *yerba mate* products can show a high 'diversification degree' depending on harvesting/cultivation methods, processing systems, packaging choices, and the existence of national or transnational supply chains in South America and worldwide. In this ambit, the concept of authenticity is a basic pillar, and the same thing can be affirmed when speaking of traceability and 'economic motivated adulteration' issues. With reference to *yerba mate*, adulteration episodes and related researches do not show worrying trends in the last decade. However, a deep analysis of selected scientific references and official Notifications in the European Union can highlight these possible actions: undeclared and fraudulent addition of carbohydrates; fraudulent identification of origin concerning *yerba mate* products; unmentioned addition and mixing of *yerba mate* with other *Ilex* species; fraudulent labelling in terms of mandatory and facultative information to the consumer. This chapter explores briefly possible food frauds in the *yerba mate* sector, with some possible countermeasure.

Keywords Addition · Authenticity · Economic motivated adulteration · Global commerce · Mixing · Origin · *Yerba mate*

Abbreviations

ATR-FTIR	Attenuated total reflectance Fourier transform infrared spectroscopy
DOU	*Diário Oficial da União*
EMA	Economic motivated adulteration
EFSA	European Food Safety Authority
EU	European Union
FNCF	Federazione Nazionale dei Chimici e dei Fisici
FBO	Food business operator
HCA	Hierarchical cluster analysis
ICP-MS	Inductively coupled plasma mass spectrometry
ICP-OES	Inductively coupled plasma optical emission spectrometry
INYM	*Instituto Nacional de la Yerba Mate*

MSME   Ministry of Micro, Small & Medium Enterprises
NFCU   National Food Crime Unit
NIR   Near-infrared spectroscopy
PCA   Principal component analysis
RASFF   Rapid Alert System for Food and Feed

8.1 Food Frauds and *Yerba Mate*. Possible Connections

The recent attribution, in Argentina, of the geographical indication for '*yerba mate*' in 2016 allows the *Instituto Nacional de la Yerba Mate* (INYM) ruling the matter of *yerba mate* commerce in terms of:

(1) Legal recognition of the product and all possible and correlated typologies and qualities
(2) Commercial classification
(3) Available information related to the origin of *yerba mate* (introducing the concept of traceability and authenticity in this field)
(4) General promotion of *yerba mate* in Argentina and worldwide (Lawson 2009; Ministerio de Agroindustria 2016).

It has to be taken into account that *yerba mate* products can show a high 'diversification degree' depending on:

(a) The harvesting and/or cultivation and harvesting method
(b) Processing systems
(c) Packaging choices
(d) The existence of national or transnational supply chains (the last option concerns mainly Uruguay and also some non-South American countries such as Syria).

In this ambit, the concept of authenticity is a basic pillar, and the same thing can be affirmed when speaking of traceability issues. Food authenticity is a product-related and intrinsic value, and it can be fully confirmed on condition that adequate and reliable traceability information are available (Parisi 2020a). In other words, the complete food authenticity is the total correspondence (total match) between claimed and declared information concerning a food product and real features of the same food product. On the contrary, the difference between claimed and real food features has to be considered in terms of food fraud, or better 'economic motivated adulteration' (EMA) (Everstine 2017; Kennedy 2012; Tucker 2011).

The basic concept of intentional adulteration or EMA (Everstine et al. 2013; Tucker 2011) should imply that one food business operator (FBO) at least in the food supply chain:

(1) Knows the difference between claimed and real food features

(2) Is willing to take advantage from the fraudulent exposition of the food product naming it in a non-real way, or claiming one or some specific features this food cannot hold
(3) Is well aware that the modification and/or omission of certain information can give some economic gain.

It has to be recognised that each FBO interested in food fraud gains is not interested in safety and health-related consequences of EMA activities. Arguments, such as health properties linked with vegetable polyphenols, traceability, quality management systems, history and social diffusion, and so on (Barbieri et al. 2014, 2019; Bhagat et al. 2019; Burris et al. 2012; Chammem et al. 2018; da Veiga et al. 2018; Delgado et al. 2016a, b, 2017, 2019; Del Águila 2014; Fiorino et al. 2019; Haddad et al. 2020a, b; Keller and Giberti 2011; Loria et al. 2009; Martin et al. 2013; Messina et al. 2015; Mania et al. 2016a, b, 2018; Parisi 2019; Parisi et al. 2020; Piovezan-Borges et al. 2016), may be discussed in this ambit. The real and appreciable result of EMA actions is only the economic advantage. On the other side, EMA may be the cause of safety incidents. Several situations—addition of urea to wheat flour, USA, 1970s; addition of melamine to milk, China, 2007 (Tucker 2011)—have demonstrated the possible occurrence of serious risks originating from EMA episodes. In other situations, the ethical and implicit agreement between producer and food consumer has been broken; the 'horse meat scandal' (addition or replacement of beef meat in beefburgers with undeclared horse meat or pig meat (United Kingdom 2007) has demonstrated the need for a reliable and powerful control system when speaking of foods and beverages in general (Meikle and McDonald 2013).

Consequently, food frauds can be considered as real crimes against the food consumers. With relation to the United Kingdom, the recent institution of the Food Standard Agency's National Food Crime Unit (NFCU) aims at preventing all possible 'food crimes', including also (Food Standards Agency 2020):

(a) Illegal processing
(b) Re-use of food, drink, or feed in the supply chain in an illegal way
(c) Adulteration: addition or replacement of undeclared ingredient(s) of lower quality and/or unacceptable types with concern to consumer safety and health
(d) Misrepresentation: commercial claims which intend define the food product with high and non-real quality
(e) False documentation supporting illegal foods.

The analysis of these situations includes also the needed countermeasures:

(1) Analysis of labelling, supporting documents, and commercial information, also by means of web-based databases and official systems for the record of Notifications and alerts (Parisi et al. 2016)
(2) Analytical detection of food frauds, with a sampling method able to define reliably a lot of food products as matching the declared information
(3) Control of information flows in the raw materials—final product direction
(4) Control of the information flow in the final product—raw materials direction

(5)   Control of information concerning product expiration in terms of sell-by-date or expiration dates (generally, the 'shelf life' information).

Many of these instruments are based on reliable 'traceability' systems. Traceability issues are certainly interesting when speaking of safety risks and correlated analyses at least, including shelf life information, according to the Parisi's First Law of Food Degradation[1] (FNCF 2020; Parisi 2002a, b, 2003, 2004, 2012, 2013, 2016; Srivastava 2019; Volpe et al. 2015). On the other hand, it has to be always remembered that authenticity concerns are strictly related to food frauds and related gains.

As discussed in Chap. 1, *yerba mate* adulteration episodes and related researches do not show worrying trends in the last decade. However, and based on the European Union (EU) Rapid Alert System for Food and Feed (RASFF) Consumers' Portal, publicly available on the web (https://webgate.ec.europa.eu/rasff-window/consumers/), we can signal that the possibility of *yerba mate*-related frauds can exist … and with a certain variety (European Commission 2020a, b).

Consequently, there is need of accurate and reliable traceability information concerning this product, also because of the possible packaging and labelling in non-South American countries. A deep analysis of selected scientific references and above-discussed RASFF Notifications can give the following directions (Crighton et al. 2019; Kucharska-Ambrożej and Karpinska 2020; Lima 2019; Marcelo et al. 2014; Porcari et al. 2016; Poswal et al. 2019; Preti 2019; Santos et al. 2020; Schneider 2017; Schneider et al. 2018; Sniechowski and Paul 2008; Trentanni Hansen et al. 2019; Vieira et al. 2020):

(1)   Undeclared and fraudulent addition of carbohydrates: saccharose, glucose, and fructose. Reason: sensorial enhancement for expired or low-quality *yerba mate* products.
(2)   Fraudulent identification of origin concerning *yerba mate* products. Reason: economic gain obtained by claiming incorrect geographical origin for *yerba mate*
(3)   Fraudulent addition and mixing of *I. paraguariensis* with other *Ilex* species. Reasons; economic gain because of the addition of cheap raw materials
(4)   Fraudulent labelling in terms of mandatory and facultative information to the consumer.

These options rely on the modification or misrepresentation of one or more of the following information (because food authenticity and its contrary, food fraud, are always linked to information):

(1)   Name of the food
(2)   Brand
(3)   Product identification
(4)   Lot identification
(5)   List of ingredients

---

[1]This Law states that 'There are not foods which are not subjected over time to a progressive transformation of their chemical, physical, organoleptic, microbiological, and structural features.

(6)   Net weight
(7)   Best-before date
(8)   Data concerning the producer, including the food registration.
(9)   Packaging date
(10)   Instructions for storage
(11)   Instruction for use
(12)   Advertising information.

The next sections explore briefly possible food frauds in the *yerba mate* sector, with some possible countermeasure.

8.2   Undeclared and Fraudulent Addition of Carbohydrates

The fraudulent addition of carbohydrates in *yerba mate* products is basically limited to sugars only. In detail, the interested sugars are as follows: saccharose, glucose, and fructose. From the legal angle, the fraudulent addition is considered irregular because of the qualitative presence, at least in Brazil, while the amount of total and separated sugars is not important enough. With relation to simple sugar addition, the product '*yerba mate* with saccharose' is allowed in Brazil at least (Ministério da Saúde 2005; Schneider 2017). Naturally, this quality should be considered as cheap enough if compared with real *yerba mate*. By the viewpoint of EMA 'players', the addition of sugars is quantitatively important, while the simple presence of undeclared sugars is sufficient to define *yerba mate* products as 'adulterated'(Schneider et al. 2018). The basic reason is to mask the unacceptable or simply low-quality features of *yerba mate* ascribed to flavour and colour. In addition, simple sugars such as glucose, fructose, and saccharose are cheap enough: their presence in the product cannot increase the real production cost, while the final price can be imposed with enhanced values on the market. The obvious weight augment is a 'plus' point in this situation. In fact, n grams of *yerba mate* leaves $+ m$ grams of saccharose $= q$ grams of *yerba mate* product, and $q > n$. The fraudulent yield is q/n: the higher the added sugar, the higher the amount or marketed *yerba mate*… with the same price of regular and legal product!

It has to be noted that the quantity of sugars in *yerba mate* has to decrease during processing (blanching). As a result, a part of lost sugars is re-added to *yerba mate* (the normal process operates in the opposite direction with the augment of polyphenols and anti-oxidants) (Dartora et al. 2011).

The fraudulent addition of sugars may pose serious health and safety risks in *yerba mate* similarly to other products (Diniz et al. 2014; Manning and Soon 2014):

(1)   Moisture can increase because sugars are easily fermentescible, with correlated diminution of shelf life values. There is naturally a theoretical risk in terms of microbial spreading

(2) Diabetics can suffer when speaking of adulterated *yerba mate* with sugar addition (because the fraudulent addition is not declared and consequently unforeseen).

By the analytical angle, the use of attenuated total reflectance Fourier transform infrared spectroscopy (ATR-FTIR) technique with associated analysis of data by means of chemometric approaches—pattern recognition: hierarchical cluster analysis (HCA) and principal component analysis (PCA)—has been proposed when speaking of the possible recognition of sugar-added *yerba mate* products with good results (Lima 2019; Schneider et al. 2018).

8.3   Fraudulent Identification of Origin Concerning *Yerba Mate* Products

The desired result, in terms of origin-related fraud, is the increase of market prices in spite of the real origin of *yerba mate* products. The fraud may concern not only the mention of a national identity (basically: Argentina, Brazil, Paraguay and Uruguay), but also the mention of a peculiar region or area into the same country.

The origin of *yerba mate* products can be difficultly defined on the basis of analytical data. Some studies claim that the use of inductively coupled plasma optical emission spectrometry (ICP-OES) and inductively coupled plasma mass spectrometry (ICP-MS) with a chemometric approach can discriminate *yerba mate* products of different geographical origins on the basis of the composition of macro- and micronutrients (metals) (Marcelo et al. 2014). Hypothetically, near-infrared spectroscopy (NIR) may be also used when speaking of the recognition of selected molecules such as methylxanthines in *yerba mate*, with the assistance of chemometric approaches (Mazur et al. 2014). Naturally, methylxanthines cannot be easily added to *yerba mate*, but their absolute amount is predicted to decrease if foreign contaminants or adulterants are present in the product. Also, this system may recognise regional or national patterns with reference to the possible identification of *yerba mate* origin (Marcelo et al. 2014).

8.4   Fraudulent Addition and Mixing of *I. Paraguariensis* with Other *Ilex* Species

The addition or mixture of non-*yerba mate* leaves to real *yerba mate* leaves would ameliorate the taste of the original raw material (in terms of reduced amount of caffeine and different quantities of saponins). In addition, claimed pharmaceutical properties may be considered in this ambit. However, the increase of market prices (in spite of the real origin of *yerba mate* products) is the undeclared and desired objective of the EMA action. Moreover, the augment of market amounts is balanced with the entering on the market of low-quality and equally priced *yerba mate*, with resulting low prices for authentic products (Haddad and Parisi 2020a, b; Parisi and Haddad 2019).

The fraudulent addition (better: mixture) of non-*yerba mate* leaves to *I. paraguariensis* can be considered as a good adulteration example. In this ambit, it has been reported that NIR technology may be used when speaking of the recognition not only of selected molecules such as methylxanthines in *yerba mate*, but also of a general spectral profile of the product, with the necessary assistance of chemometric approaches (Mazur et al. 2014). This system may be able to recognise regional/national patterns, and also abnormal spectral profiles belonging to non-*I. paraguariensis* products because of the addition (or substitution) of other *Ilex* species. An hypothetical example can be the use of *I. argentina* (found in the area approximately located between Santa Cruz de la Sierra, Bolivia, and Andagalá, Argentina) which has been often confused in the past with *I. paraguariensis* (for this reason, *yerba mate* was considered a native Bolivian plant, until recent times) (Giberti 1995).

8.5 Fraudulent Labelling in Terms of Mandatory and Facultative Information to the Consumer

This section concerns all possible mislabelling and undeclared information with legal importance, when speaking of *yerba mate*. One of these information—the indication of 'best-before use'—is extremely important because it is related to the possible addition of sugars. As above mentioned, moisture can increase with reduction of shelf life values. On the other hand, the best-before end date informs the consumers that the products can be still consumed after a certain date (no expiration date), even if the food is adulterated and substantially is not equal to a normal *yerba mate* product! In other terms, a non-*yerba mate* genuine product can really expire (and be dangerous) after a defined date instead of normal *yerba mate* (which a specific danger is not recognised after a specific date)… because the product is named '*yerba mate*'.

The problem is that 'shelf life' may be considered as a 'best-before end' date, with a clear meaning of commercial acceptability (or unacceptability) instead of safety and health acceptability. Available data in the literature show shelf life values up to 30–44 or 10–13 days (depending on storage conditions, relative humidity above all) (dos Santos et al. 2020; Surkan et al. 2009). It has also been mentioned that the most reliable dates for *yerba mate* should be the harvesting and the packaging date. Actually, shelf life is three years after packaging, in Argentina and Paraguay at least (Goyerbamate.com 2020). As a result, the date of packaging should define the real durability of the product. On the other hand, some products mention clearly the 'expiration date' information on their packages on the market… and this definition might not imply commercial acceptability without safety and health features. The recent 'Guidance on date marking and related food information: part 1 (date marking)' by the European Food Safety Authority (EFSA) clearly demonstrates that 'best-before end' is not the same thing of 'use-by date'… (EFSA Panel on Biological Hazards 2020). In addition, the printed date can be printed as 'month/year' or 'day/month/year' (even if the order of numbers is different in South America). These two systems are generally ascribed to 'best-before date' and 'use-by date' information. The concomitant existence of *yerba mate* products with both information types can be questionable, and food consumers may be confused.

This argument does not mention all possible products containing *yerba mate*, such as alcoholic beverages and espresso-like pods etc.

Other mislabelling options such as incorrect or modified shelf life dates can be observed, and the possible countermeasures rely on a reliable flow if input and output data (the 'traceability' system). At present, more research is surely needed in the ambit of *yerba mate* products, also taking into account the international ambit of related commerces, and the expansion of electronic (on-line) businesses concerning *yerba mate*. The market of electronic transactions is extremely rapid and the number of possible adulteration episodes can increase!

References

Barbieri G, Barone C, Bhagat A, Caruso G, Conley Z, Parisi S (2014) The prediction of shelf life values in function of the chemical composition in soft cheeses. In: The influence of chemistry on new foods and traditional products. Springer International Publishing, Cham. https://doi.org/10.1007/978-3-319-11358-6_2

Barbieri G, Bergamaschi M, Saccani G, Caruso G, Santangelo A, Tulumello T, Vibhute B, Barbieri G (2019) Processed Meat and Polyphenols: Opportunities, Advantages, and Difficulties. J AOAC Int 102(5):1401–1406. https://doi.org/10.1093/jaoac/102.5.1401

Bhagat AR, Delgado AM, Issaoui M, Chammem N, Fiorino M, Pellerito A, Natalello S (2019) Review of the role of fluid dairy in delivery of polyphenolic compounds in the diet: chocolate milk, coffee beverages, Matcha green tea, and beyond. J AOAC Int 102(5):1365–1372. https://doi.org/10.1093/jaoac/102.5.1365

Burris KP, Harte FM, Davidson PM, Stewart Jr CN, Zivanovic S (2012) Composition and bioactive properties of yerba mate (Ilex paraguariensis A. St.-Hil.): a review. Chil J Agric Res 72, 2:268–274

Chammem N, Issaoui M, De Almeida AID, Delgado AM (2018) Food crises and food safety incidents in European Union, United States, and Maghreb Area: current risk communication strategies and new approaches. J AOAC Int 101(4):923–938. https://doi.org/10.5740/jaoacint.17-0446

Crighton E, Coghlan ML, Farrington R, Hoban CL, Power MW, Nash C, Mullaney I, Byard RW, Trengove R, Musgrave IF, Bunce M, Maker G (2019) Toxicological screening and DNA sequencing detects contamination and adulteration in regulated herbal medicines and supplements for diet, weight loss and cardiovascular health. J Pharm Biomed Anal 176:112834. https://doi.org/10.1016/j.jpba.2019.112834

da Veiga DTA, Bringhenti R, Copes R, Tatsch E, Moresco RN, Comim FV, Premaor MO (2018) Protective effect of yerba mate intake on the cardiovascular system: a post hoc analysis study in postmenopausal women. Braz J Med Biol Res 51(6):e7253. https://doi.org/10.1590/1414-431x20187253

Dartora N, de Souza LM, Santana AP, Iacomini M, Valduga AT, Gorin PAJ, Sassaki GL (2011) UPLC-PDA–MS evaluation of bioactive compounds from leaves of Ilex paraguariensis with different growth conditions, treatments and ageing. Food Chem 129(4):1453–1461. https://doi. org/10.1016/j.foodchem.2011.05.112

Del Águila Á (2014) A través de la yerba mate: etnicidad y racionalidad económica entre los trabajadores rurales paraguayos en la industria de la construcción de Buenos Aires. Antípoda. Rev Antropol Arqueol 18:165–187. https://doi.org/10.7440/antipoda18.2014.08

Delgado AM, Almeida MDV, Parisi S (2017) Chemistry of the Mediterranean Diet. Springer International Publishing, Cham. https://doi.org/10.1007/978-3-319-29370-7

Delgado AM, Issaoui M, Chammem N (2019) Analysis of main and healthy phenolic compounds in foods. J AOAC Int 102(5):1356–1364. https://doi.org/10.1093/jaoac/102.5.1356

Delgado AM, Vaz de Almeida MD, Barone C, Parisi S (2016) Leguminosas na Dieta Mediter-rânica—Nutrição, Segurança, Sustentabilidade. CISA—VIII Conferência de Inovação e Segurança Alimentar, ESTM- IPLeiria, Peniche, Portugal

Delgado AM, Vaz de Almeida MD, Parisi S (2016a) Chemistry of the mediterranean diet. Springer International Publishing, Cham. https://doi.org/10.1007/978-3-319-29370-7

Diniz PHGD, Gomes AA, Pistonesi MF, Band BSF, de Araújo MCU (2014) Simultaneous classi-fication of teas according to their varieties and geographical origins by using NIR spectroscopy and SPALDA. Food Anal Methods 7:1712–1718. https://doi.org/10.1007/s12161-014-9809-7

dos Santos LF, Vargas BK, Bertol CD, Biduski B, Bertolin TE, dos Santos LR, Brião VB (2020) Clarification and concentration of yerba mate extract by membrane technology to increase shelf life. Food Bioproducts Proc: Trans Inst Chem Eng Part C 122:22–30

EFSA Panel on Biological Hazards (2020) Guidance on date marking and related food information: part 1 (date marking). European Food Safety Authority (EFSA), Parma. Available https://www. efsa.europa.eu/en/efsajournal/pub/6306. Accessed 15 Dec 2020

European Commission (2020a) RASFF Portal. Notification details—2019.2199. Avail-able https://webgate.ec.europa.eu/rasff-window/portal/?event=notificationDetail&NOTIF_REF ERENCE=2019.2199. Accessed 15 Dec 2020

European Commission (2020b) RASFF Portal. Notification details - 2020.0380. Avail-able https://webgate.ec.europa.eu/rasff-window/portal/?event=notificationDetail&NOTIF_REF ERENCE=2020.0380. Accessed 15 Dec 2020

Everstine K (2017) Supply chain complexity and economically motivated adulteration. In: Kennedy S (ed) (2016) Food protection and security: preventing and mitigating contamination during food processing and production. Woodhead Publishing Series in Food Science, Technology and Nutrition. Woodhead Publishing, Duxford, Cambridge, and Kidlington, pp 1–14

Everstine K, Spink J, Kennedy S (2013) Economically motivated adulteration (EMA) of food: common characteristics of EMA incidents. J Food Prot 76(4):723–725. https://doi.org/10.4315/ 0362-028X.JFP-12-399

Fiorino M, Barone C, Barone M, Mason M, Bhagat A (2019) The intentional adulteration in foods and quality management systems: chemical aspects. Quality systems in the food industry. Springer International Publishing, Cham, pp 29–37

FNCF (2020) La Prima Legge della degradazione a-limenti prende il nome da un Chimico italiano. Federazione Nazionale dei Chimici e dei Fisici (FNCF), Roma. Available https://www.chimic ifisici.it/la-prima-legge-della-degradazione-alimenti-prende-il-nome-da-un-chimico-italiano/. Accessed 11 Dec 2020

Food Standards Agency (2020) Food crime. Understanding food crime and how to report it. Food Standards Agency, London. Available https://www.food.gov.uk/safety-hygiene/food-crime. Accessed 15 Dec 2020

Giberti GC (1995) Aspectos oscuros de la corología de Ilex paraguariensis St. Hil. In: Winge H, Ferreira AG, de Araujo Mariath JE, Tarasconi LC (eds) "Erva-Mate: biologia e cultura no Cone Sul. Editora da UFRGS, Porto Alegre, pp 289–300

Goyerbamate.com (2020) Expired Yerba Mate? Goyerbamate.com. Available https://www.goy erbamate.com/Expired-Yerba-Mate_c_46.html#:~:text=The%20first%20point%20we%20n eed,make%20the%20yerba%20age%20faster. Accessed 15 Dec 2020

Haddad MA, Dmour H, Al-Khazaleh JFM, Obeidat M, Al-Abbadi A, Al-Shadaideh AN, Almazra'awi MS, Shatnawi MA, Iommi C (2020a) Herbs and medicinal plants in Jordan. J AOAC Int 103(4):925–929. https://doi.org/10.1093/jaocint/qsz026

Haddad MA, El-Qudah J, Abu-Romman S, Obeidat M, Iommi C, Jaradat DSM (2020b) Phenolics in mediterranean and middle east important fruits. J AOAC Int 103(4):930–934. https://doi.org/10.1093/jaocint/qsz027

Haddad MA, Parisi S (2020a) Evolutive profiles of Mozzarella and vegan cheese during shelf-life. Dairy Industries International 85(3):36–38

Haddad MA, Parisi S (2020b) The next big HITS. New Food Magazine 23(2):4

Keller HA, Giberti GC (2011) Primer registro para la flora argentina de Ilex affinis (Aquifoliaceae), sustituto de la 'yerba mate'. Boletín de la Sociedad Argentina de Botánica 46(1–2):187–194

Kennedy S (2012) Emerging global food system risks and potential solutions. In: Ellefson W, Zach L, Sullivan D (eds) Improving import food safety. Wiley Inc. Hoboken, pp 1–20. https://doi.org/10.1002/9781118464298.ch1

Kucharska-Ambrożej K, Karpinska J (2020) The application of spectroscopic techniques in combination with chemometrics for detection adulteration of some herbs and spices. Microchem J 153:104278. https://doi.org/10.1016/2Fj.microc.2019.104278

Lawson J (2009) Cultivating green gold: a political ecology of land use changes for small yerba mate farmers in misiones, Argentina. Dissertation, Yale School of Forestry, New Haven, CT

Lima PCD (2019) Discriminação de erva-mate para chimarrão quanto à origem geográfica e presença de açúcar utilizando FTIR e quimiometria. Dissertation, Universidade Tecnológica Federal do Paraná, Curitiba

Loria D, Barrios E, Zanetti R (2009) Cancer and yerba mate consumption: a review of possible associations. Rev Panam Salud Púb 25:530–539

Mania I, Barone C, Caruso G, Delgado A, Micali M, Parisi S (2016a) Traceability in the cheese-making field. The Regulatory Ambit and Practical Solutions. Food Qual Mag 3:18–20. ISSN 2336-4602

Mania I, Delgado AM, Barone C, Parisi S (2018) Traceability in the dairy industry in Europe. Springer International Publishing, Heidelberg, Germany

Mania I, Fiorino M, Barone C, Barone M, Parisi S (2016b) Traceability of packaging materials in the cheesemaking field. The EU Regulatory Ambit. Food Packag Bull 25, 4&5:11–16

Manning L, Soon JM (2014) Developing systems to control food adulteration. Food Pol 49(1):23–32. https://doi.org/10.1016/j.foodpol.2014.06.005

Marcelo MCA, Martins CA, Pozebon D, Ferrão MF (2014) Methods of multivariate analysis of NIR reflectance spectra for classification of yerba mate. Anal Methods 6(19):7621–7627. https://doi.org/10.1039/C4AY01350F

Martin JGP, Porto E, de Alencar SM, da Glória EM, Corrêa CB, Cabral ISR (2013) Antimicrobial activity of yerba mate (Ilex paraguariensis St. Hil.) against food pathogens. Rev Arg Microbiol 45, 2:93–98. https://doi.org/10.1016/s0325-7541(13)70006-3

Mazur L, Peralta-Zamora PG, Demczuk B, Hoffmann-Ribani R (2014) Application of multivariate calibration and NIR spectroscopy for the quantification of methylxanthines in yerba mate (Ilex paraguariensis). J Food Compos Anal 35(2):55–60. https://doi.org/10.1016/j.jfca.2014.04.005

Meikle J, McDonald H (2013) Cameron tells supermarkets: horsemeat burger scandal unacceptable. The guardian, London. Available https://www.theguardian.com/world/2013/jan/16/tesco-burgers-off-shelves-horsemeat. Accessed 15 Dec 2020

Messina D, Soto C, Méndez A, Corte C, Kemnitz M, Avena V, Dal Balzo D, Pérez Elizalde R (2015) Efecto hipolipemiante del consumo de mate en individuos dislipidémicos. Nutr Hosp 31(5):2131–2139

Ministério da Saúde (2005) RDC n° 277 de 22 de setembro de 2005. Diário Oficial da União (DOU) of 23rd September 2005 N° 184, sexta-feira:379–380

Ministerio de Agroindustria (2016) Resolución N° 13/2016—Aprueba el Protocolo de Producción, Elaboración y/o Guía de Prácticas de la 'Yerba Mate Argentina'. Boletín Oficial N° 33337, 15 de marzo de 2016, p 31. Available http://www.fao.org/faolex/results/details/en/c/LEX-FAOC15 3569. Accessed 15 Dec 2020

Parisi S (2002a) I fondamenti del calcolo della data di scadenza degli alimenti: principi ed applicazioni. Ind Aliment 41, 417:905–919

Parisi S (2002b) Profili evolutivi dei contenuti batterici e chimico-fisici in prodotti lattiero-caseari. Ind Aliment 41(412):295–306

Parisi S (2003) Evoluzione chimico-fisica e microbiologica nella conservazione di prodotti lattiero-caseari. Ind Aliment 42(423):249–259

Parisi S (2004) The prediction of Shelf Life about cheese on the basis of storage temperature. Italian J food Sci Special Issue, pp 11–19

Parisi S (2012) Food packaging and food alterations. The User-oriented Approach, Smithers Rapra Technology Ltd, Shawbury

Parisi S (2013) Food industry and packaging materials—performance-oriented guidelines for users. Smithers Rapra Technologies, Shawsbury

Parisi S (2016) The world of foods and beverages today: globalization, crisis management and future perspectives. Learning.ly/ The Economist Group, available http://learning.ly/products/the-world-of-foods-and-beverages-today-globalization-crisis-management-and-future-perspectives. Accessed 9 Dec 2020

Parisi S (2019) Analysis of major phenolic compounds in foods and their health effects. AOAC J 102(5):1354–1355. https://doi.org/10.5740/jaoacint.19-0127

Parisi S (2020a) Course: fundamentals of food traceability. Lourdes Matha Institute of Hotel Management and Catering Technology. First Lecture: 14 Dec 2020

Parisi S, Barone C, Sharma RK (2016) Chemistry and food safety in the EU.The Rapid Alert System for Food and Feed (RASFF). Springer Briefs in Molecular Science, SpringerInternational Publishing, Cham

Parisi S, Dongo D, Parisi C (2020) Resveratrolo, conoscenze attuali e prospettive. Great Italian Food Trade 27/10/2020. Available www.greatitalianfoodtrade.it/integratori/resveratrolo-conoscenze-attuali-e-prospettive. Accessed 9 Dec 2020

Parisi S, Haddad MA (2019) Food safety 101. Al-Balqa Applied University, Al-Salt, Jordan

Piovezan-Borges AC, Valério-Júnior C, Gonçalves IL, Mielniczki-Pereira AA, Valduga AT (2016) Antioxidant potential of yerba mate (Ilex paraguariensis St. Hil.) extracts in Saccharomyces cerevisae deficient in oxidant defense genes. Braz J Biol 76(2):539–544. https://doi.org/10.1590/1519-6984.01115

Porcari AM, Fernandes GD, Barrera-Arellano D, Eberlin MN, Alberici RM (2016) Food quality and authenticity screening via easy ambient sonic-spray ionization mass spectrometry. Anal 141(4):1172–1184. https://doi.org/10.1039/C5AN01415H

Poswal FS, Russell G, Mackonochie M, MacLennan E, Adukwu EC, Rolfe V (2019) Herbal teas and their health benefits: a scoping review. Plant Foods Hum Nutr 74:266–276. https://doi.org/10.1007/s11130-019-00750-w

Preti R (2019) Progress in Beverages authentication by the application of analytical techniques and chemometrics. In: Grumezescu AM, Holban AM (eds) Quality control in the beverage industry, vol 17: the Science of Beverages. Academic Press, Sawston, pp 85–121. https://doi.org/10.1016/b978-0-12-816681-9.00003-5

Santos MCD, Azcarate SM, Lima KMG, Goicoechea HC (2020) Fluorescence spectroscopy application for Argentinean yerba mate (Ilex paraguariensis) classification assessing first-and-second-order data structure properties. Microchem J 155:104783. https://doi.org/10.1016/j.microc.2020.104783

Schneider M (2017) Determinação da adulteração da erva-mate por adição de sacarose empregando espectroscopia no infravermelho (atr-ftir) em conjunto com ferramentas quimiométricas. Dissertation, Universidade Federal do Rio Grande do Sul, Porto Alegre

Schneider M, Schneider RC, Corbellini VA, Mahlmann CM, Fior CS, Ferrão MF (2018) Exploratory analysis applied for the evaluation of yerba mate adulteration (Ilex paraguariensis). Food Anal Methods 11(7):2035–2041. https://doi.org/10.1007/s12161-018-1202-5

Sniechowski VI, Paul LM (2008) The labeling on the" yerba mate"(Ilex Paraguariensis) packages in the mercosur"(South American Common Market). Visión de futuro 9(1):105–124

Srivastava PK (2019) Status report on bee keeping and honey processing. Status report on bee keeping and honey processing. Development Institute, Ministry of Micro, Small & Medium Enterprises (MSME), Government of India 107, Industrial Estate, Kalpi Road, Kanpur-208012. Available   http://msmedikanpur.gov.in/cmdatahien/reports/diffIndustries/Status%20Report%20on%20Bee%20keeping%20&%20Honey%20Processing%202019-2020.pdf. Accessed 9 Dec 2020

Surkan S, Albani O, Ramallo L (2009) Influence of storage conditions on sensory shelf life of yerba mate. J Food Qual 32(1):58–72. https://doi.org/10.1111/j.1745-4557.2008.00236.x

Trentanni Hansen GJ, Almonacid J, Albertengo L, Rodriguez MS, Di Anibal C, Delrieux C (2019) NIR-based Sudan I to IV and Para-Red food adulterants screening. Food Addit Contam Part A 36(8):1163–1172. https://doi.org/10.1080/19440049.2019.1619940

Tucker J (2011) Economically motivated adulteration. Case studies in agricultural biosecurity, 1. Agroterrorism and food safety. Federation of American Scientists, Wagshington. Available https://fas.org/biosecurity/education/dualuse-agriculture/1.-agroterrorism-and-foodsafety/economically-motivated-adulteration.html. Accessed 15 Dec 2020

Vieira TF, Makimori GYF, dos Santos Scholz MB, Zielinski AAF, Bona E (2020) Chemometric Approach Using ComDim and PLS-DA for Discrimination and Classification of Commercial Yerba Mate (Ilex paraguariensis St. Hil.). Food Anal Methods 13(1):97–107. https://doi.org/10.1007/s12161-019-01520-9

Volpe MG, Di Stasio M, Paolucci M, Moccia S (2015) Polymers for food shelf-life extension. Functional polymers in food science, Scrivener Publishing LLC. pp 9–66

Printed in the United States
by Baker & Taylor Publisher Services